PAUL JOHN FLORY

A Life of Science and Friends

PAUL JOHN FLORY

A Life of Science and Friends

Gary D. Patterson
Carnegie Mellon University
Pittsburgh, PA, USA

James E. Mark
University of Cincinnati
Cincinnati, OH, USA

Joel R. Fried
University of Louisville
Louisville, KY, USA

Do Y. Yoon
Stanford University
Stanford, CA, USA

CRC Press is an imprint of the
Taylor & Francis Group, an **informa** business

CRC Press
Taylor & Francis Group
6000 Broken Sound Parkway NW, Suite 300
Boca Raton, FL 33487-2742

Printed on acid-free paper
Version Date: 20150224

International Standard Book Number-13: 978-1-4665-9576-7 (Paperback)

Library of Congress Cataloging-in-Publication Data

Paul John Flory : a life of science and friends / Gary D. Patterson, James E. Mark, Joel R. Fried, Do Y. Yoon.
pages cm
"A CRC title."
Includes bibliographical references and index.
ISBN 978-1-4665-9576-7 (alk. paper)
1. Flory, Paul J. 2. Chemists--United States--Biography. 3. Nobel Prize winners--United States--Biography. I. Patterson, Gary D. (Gary David), 1946- II. Mark, James E., 1934- III. Fried, Joel R. IV. Yoon, D. Y.
QD22.F57J64 2016

540.92--dc23
[B] 2015006720

Visit the Taylor & Francis Web site at
http://www.taylorandfrancis.com

and the CRC Press Web site at
http://www.crcpress.com

Contents

List of Figures

Acknowledgments

This project was initiated during a fellowship year at the Chemical Heritage Foundation in 2004–2005. There are 72 boxes of Flory materials in their archives. The project has been encouraged ever since by James Mark of the University of Cincinnati.

The biography of the Flory family has been greatly assisted by Susan Flory Springer. The Flory family tree has been traced by her efforts back to 1728. She has also provided background on both her mother and father, as well as other family pictures and stories.

One of the most significant places in the life of Paul John Flory was Manchester College (now University). The Brethren Historical Collection and its archivist, Jeanine Wine, have made possible the richly illustrated family history. The Manchester University Archives generously provided both images and documentary materials for the chapter on Flory at Manchester College. The complete collection of Paul Flory's medals and awards are housed in the university library.

The Ohio State University Archives were also a rich supply of images and documents. Paul Flory is well remembered at Ohio State, both as a famous scientist and as a distinguished human being.

The chapter on Flory at Cincinnati was written by Joel Fried. The University of Cincinnati Archives provided significant material both on the history of the Basic Sciences Research Laboratory and on the interaction of Paul Flory with the University of Cincinnati.

Roberta Rehner Iverson graciously provided pictures and background material on her father, John Rehner, Jr. Background on the Esso Laboratories was obtained from the official corporate history: *New Horizons: 1927–1950* by Larson, Knowlton, and Popple, Harper and Row, New York, 1971.

The chapter on Goodyear was researched at the company archives, housed at the University of Akron. Many thanks are due to the helpful staff.

I wish to thank James Peters from the University of Manchester (UK) Archives for ably preparing the material on Paul Flory during my visit there.

The Carnegie Mellon University Archives and the Mellon Institute Library were helpful in preparing the chapter on Flory's work at the Mellon Institute.

I wish to thank Grace Baysinger of the Stanford chemistry department library for substantial help in obtaining documents and photographs from the Stanford period. I also wish to acknowledge the archives of the Hoover Institution at Stanford University for access to the special collection on Flory's humanitarian activities.

Since Paul Flory was a world statesman as well as a scientist, it was necessary to visit archives in many places. Part of the joy of preparing this biography has been to see in person where he worked and lived.

I also wish to thank the many friends of Paul Flory for contributing personal accounts of their interactions with him. A much fuller picture of Paul Flory as a man has resulted.

Extensive archives of Paul Flory's humanitarian work are housed in the Hoover Institution at Stanford University. In addition, substantial material is available at the Chemical Heritage Foundation. I wish especially to thank Andrew Mangravite of the Chemical Heritage Foundation (CHF) for his help with these archives.

About the Authors

Gary Patterson is professor of chemical physics and polymer science at Carnegie Mellon University. He obtained his BS in chemistry from Harvey Mudd College (1968). He earned his PhD in physical chemistry from Stanford University (1972) working with Paul John Flory. He was a member of the technical staff in the Chemical Physics Department of AT&T Bell Laboratories from 1972 to 1984 and received the National Academy of Sciences Award for initiatives in research in 1981 for his work on the structure and dynamics of amorphous polymers using light scattering. He is a Fellow of the American Physical Society and the Royal Society of Chemistry. He has published more than 100 papers in technical journals such as *Macromolecules*, the *Journal of Chemical Physics*, and the *Journal of Polymer Science*. He is the author of *Physical Chemistry of Macromolecules* (CRC Press, 2007). He is now a chemical historian and has published many articles and books on the history of polymer science and the history of physical chemistry. He is the chair of the History of Chemistry Division of the American Chemical Society. He has been associated with the Chemical Heritage Foundation since 2004 and is now the chair of the Heritage Council and is a member of the Board of Directors.

James E. Mark is professor of chemistry emeritus at the University of Cincinnati. He earned his BS in chemistry from Wilkes College (1957) and PhD in physical chemistry from the University of Pennsylvania (1962). He fell in love with polymer science while working with Tom Fox at Rohm and Haas during a break in his undergraduate years. He followed this star as a postdoctoral fellow with Paul Flory at Stanford. He rose to the position of professor at the University of Michigan (1972) before being appointed to the University of Cincinnati in 1977. He was the director of the Polymer Research Center and a distinguished research professor. He has published extensively in the areas of physical chemistry of macromolecules, especially rubber. He has been honored with many awards including the Charles Goodyear Medal of the Rubber Division of the American Chemical Society (ACS), the ACS Applied Polymer Science Award, and the Flory Polymer Education Award of the ACS Division of Polymer Chemistry. He is one of the editors of the *Selected Works of Paul Flory*, and has done more than anyone in contributing to the science associated with Paul Flory.

Joel R. Fried is currently professor and chair of the Department of Chemical Engineering at the University of Louisville. Prior to this appointment, he was professor and chair of the Department of Chemical and Biomedical Engineering at Florida State and was professor of Materials Engineering and the Wright Brothers Institute endowed chair in Nanomaterials at the University of Dayton. He is also professor emeritus and Fellow of the Graduate School at the University of Cincinnati where he was professor and head of the Department of Chemical Engineering and director of the Polymer Research Center. He has authored more than 150 journal articles, several patents, and book chapters and is the author of *Polymer Science and Technology* (3rd edition, Prentice-Hall). He earned his BS in biology from RPI (Rensselaer Polytechnic Institute), his BS and ME in chemical engineering also from RPI, and he earned his MS and PhD degrees in polymer science and engineering from the University of Massachusetts Amherst.

Do Y. Yoon is consulting professor of chemical engineering at Stanford University since 2012. He obtained his BS in chemical engineering from Seoul National University (1969), and earned his PhD in polymer science and engineering from the University of Massachusetts Amherst, working with Richard S. Stein (1973). He did his postdoctoral study with Paul J. Flory in the chemistry department of Stanford University (1973–1975). He then worked at the IBM Research Laboratory in San Jose as a research staff member and manager of Polymer Physics Group (1975–1999). From 1999 to 2012, he was professor of chemistry at Seoul National University. He was also visiting professor at the Max Planck Institute for Polymer Research (2000–2006), National Institute of Standards and Technology (2006–2012), and University of Bayreuth (2012). He has published about 250 research papers, was elected a Fellow of the American Physical Society in 1985, and received a Senior Humboldt Research Award in 1999. His research areas include molecular conformations, chain dynamics, semicrystalline morphology, liquid crystalline order, surface and thin film characteristics of polymers, and structure–property relationships of polymers for information technology and clean energy. He is a co-editor of the *Selected Works of Paul J. Flory*.

1

Paul John Flory: A Life of Science and Friends

Introduction

Paul John Flory (1910–1985) was one of the greatest scientists of the twentieth century. He received the Nobel Prize in chemistry in 1974 for "his fundamental achievements, both theoretical and experimental, in the physical chemistry of macromolecules."[1] His classic monograph from 1953, *Principles of Polymer Chemistry*, is still one of the essential books in the field.[2] The scientific career of Paul Flory will be presented in all its glory in this book.

But, Paul Flory will be remembered for much more than his technical contributions to polymer science, great as they were. He made a large number of scientific friends during his life and greatly influenced both their science and their lives. For those friends that have joined him in death, short biographical vignettes will be included in this volume. For those friends who are still writing in 2014, personal memorials of a longer format will be included. Both lists are long, but with the luxury of a formal biography, it is possible to do justice to this aspect of Flory's life.

Paul Flory will also be remembered as a tireless crusader for human rights. His efforts on behalf of many scientists living under Soviet rule are well documented and the story needs to be told. An extensive presentation of this aspect of his life will be presented (Chapter 17).

Final reflections on his character as a human, his significance for science, and his capacity for true friendship attempt to draw a picture of the man, Paul John Flory.

The family background of Paul John Flory is fascinating. A detailed study is included in Chapter 2. His father, Ezra Younce Flory (1870–1940), was a major factor in forming his character. Interesting details of his childhood are also presented. The advanced education of Paul Flory includes Manchester College (BS 1931) (Chapter 3) and Ohio State University (PhD 1934) (Chapter 4).

Paul Flory's early industrial career includes DuPont (1934–1938) (Chapter 5), Standard Oil Development Company (Esso) (1940–1943) (Chapter 7), and Goodyear Tire and Rubber Company (1943–1948) (Chapter 8). He also spent a brief period at the University of Cincinnati Basic Science Research Laboratory (1938–1940) (Chapter 6).

Paul Flory's academic career includes Cornell University (1948–1956) (Chapter 9), The University of Manchester (1954–1955) (Chapter 10), The Mellon Institute for Industrial Research (1956–1960) (Chapter 11), and Stanford University (1960–1985) (Chapters 12 and 13).

Within the context of these phases of his career, his science and his friends will be discussed. Many of these details have been presented in the paper "Paul John Flory: Physical Chemist and Humanitarian."[3] There are also significant details contained in the book *Polymer Science* 1935–1953: *Consolidating the Paradigm*.[4] Major previous biographical details are included in the *National Academy of Sciences Biography* by Johnson et al.[5] and in the *Dictionary of Scientific Biography* article on Paul Flory by Morris.[6]

The existence of a large number of living former colleagues of Paul Flory has made possible an extensive collection of personal remembrances (Chapter 15). The format was designed to include a brief personal narrative, the circumstances under which they worked with Flory, a discussion of their joint scientific work, and an assessment of Paul Flory's place in polymer science, chemistry, and world science.

This biography would be egregiously incomplete without a chapter on Paul Flory's soul mate and wife, Emily T. Flory (1912–2006) (Chapter 14). She supported his career wholeheartedly and taught him many things about society and philanthropy.

Final chapters paint narrative pictures of Paul Flory as a scientist (Chapter 18) and as a friend (Chapter 19).

2

History of the Flory Family

The available sources for the history of the Flory family in America are especially rich. One source is from *The Brethren Encyclopedia.*[7] Another extensive source results from another current branch of the family and its extensive research, both in Europe and the United States.[8] *A History of the German Baptist Brethren in Europe and America* by Brumbaugh[9] was very enlightening. The Flory family is well documented on the genealogical website Ancestry.com.[10] Also, the archives of Manchester University were very helpful. The Brethren Historical Collection is a part of the archives.

The focal point for the early history is an epic voyage on the ship *Hope,* from Rotterdam to Philadelphia in 1733. Good records of the passengers were kept and are extant.[10] One Joseph J. Flory (1682–1741) was a passenger on that ship which docked at Philadelphia on August 28, 1733. He was accompanied by his wife, Mary (1690–1741), and many children. They settled in Rapho Township, Lancaster County, Pennsylvania. Rapho Township in that era was largely just farms, and the current principal town of Manheim was only founded in 1762 (Figure 2.1).

Many of the passengers on the *Hope* that settled in Rapho Township were members of the German Baptist Brethren from the Palatinate in Germany. The Flory family remained faithful to this tradition for many generations. Speculation that the Flory family was from Huguenot stock[7] may not be accurate and an extensive amount of research in Switzerland suggests that they were part of the Swiss Mennonites that were driven out of Switzerland into Germany.[8] Upon settling in Pennsylvania, the Flory family joined the Conestoga Congregation of the Church of the Brethren in Leola, Pennsylvania.[7]

Soon after getting settled in America, Joseph and Mary (perhaps Anna Maria) had a son Abraham (1735–1827). Some Florys moved to Maryland and Virginia and some remained in Pennsylvania. There are Florys in Manheim to this day. But, Abraham eventually moved west, first to Brothersvalley, Somerset County, Pennsylvania from 1796 to 1809, and then to Montgomery County, Ohio, where his three sons lived. Abraham died on the Beeghly farm, the home of his son Joseph (1769–1823) and wife Elizabeth Beeghly Flory (1778–1823) in 1827. Abraham Flory and his son Emanuel (1776–1849) were ministers in Montgomery County. *The Brethren Encyclopedia* proudly notes that more than 200 ministers were among the descendants of Joseph Flory, Sr.[7]

Joseph and Elizabeth Flory had a son, Joseph E. Flory (1801–1874), who lived and died in Montgomery County, Ohio. Joseph and his wife Rosenia

FIGURE 2.1
Manheim, Rapho Township, Pennsylvania. (GDP photo.)

Bennett (1805–1883) had a son John D. Flory (1839–1919) who was born in Dayton, Ohio (Montgomery County). John and his wife Permelia (Millie) Younce (1845–1923) had a son Ezra Younce Flory (1870–1940) who was born in Union City, Miami County, Ohio. Ohio was clearly a good place for the Flory family and they thrived there among their Dunker Brethren (Figure 2.2).

Ezra Younce Flory has a listing in *The Brethren Encyclopedia* (Figure 2.3). He joined the Salem Congregation of the Church of the Brethren in 1889. After a year of premedical education, he married Emma Brumbaugh (1872–1904) in 1893 and began a career teaching within the public school. An active participant in his congregation's Sunday School, Ezra was elected a minister in 1901. His first wife died in childbirth in 1904 and he married Martha Brumbaugh (1871–1960), an alumna of Manchester College, in 1905. In 1907 he was called to become pastor of the Sterling, Illinois congregation. He was elected to the eldership in 1910 and served as the presiding Elder of

FIGURE 2.2
Henry Flory farm (Montgomery County, Ohio, 1903). (From Leonard Flory. With permission.)

FIGURE 2.3
Ezra Younce Flory (1870–1940). (From Manchester University Archives and Brethren Historical Collection. With permission.)

the Sterling Congregation from 1911 to 1913. Paul John Flory (1910–1985) was born on June 19, 1910 in Sterling, Illinois. It was a good year!

In conjunction with pastoral work, Ezra Flory embarked on a program of educational improvement, which was fostered by the founding in 1905 of *Bethany Bible School* in Chicago, Illinois. He obtained two degrees from there including a Bachelor of Divinity (1916). He polished his theological training with a Masters of Arts Degree from Manchester College (1918). Ezra Flory utilized this educational background to teach Christian education at Bethany Bible School and to become the first full-time general secretary of the General Sunday School Board from 1920 to 1928. He traveled extensively with the idea that pedagogically sound training was basic to the future of the church. He finally obtained a Doctorate in theology from the Southern Baptist Theological Seminary in Louisville, Kentucky in 1929. Paul Flory learned about dedication and hard work from his father. He also learned to value education (Figure 2.4).

The 1920 Federal Census lists Paul Flory in Chicago, the location of the Bethany Bible School. When Ezra became general secretary the Florys settled in Elgin, Illinois, the location of the denominational headquarters for the Church of the Brethren. Paul Flory graduated from Elgin High School in 1927. His older sister, Mrs. Harry Wondergem (Margaret) (1896–1985), remarked: "I had some early indication that Paul would be famous because he was always at the head of his classes in both grade and high schools."[11]

FIGURE 2.4
Bethany Bible School, Chicago, Illinois. (From Manchester University Archives and Brethren Historical Collection. With permission.)

3

Flory at Manchester College (Now University)

Manchester College in North Manchester, Indiana, seemed the perfect choice for Paul Flory, the son of a Church of the Brethren pastor in nearby Huntington, Indiana. Grounded in the values and traditions of the Church of the Brethren, Manchester has had a strong academic focus on the natural sciences since its founding in 1889. That story is beautifully told by William R. Eberly.[12] The science program at Manchester started with the biologist Albert B. Ulrey (1860–1932) (Figure 3.1), a student of Oliver P. Jenkins, later of Stanford University. He was also influenced by David Starr Jordan of Indiana University (and later president of Stanford University). Ulrey taught at Manchester from 1894 to 1901 and then was appointed at the University of Southern California. Another important figure in the history of science at Manchester was Levi Daniel Ikenberry (1867–1957). He arrived at Manchester in 1900 and was tasked with putting the college affairs in order. He did far more than that and served until 1943. Another pillar in the science program was Edward Kintner (1879–1975) (Figure 3.1). He taught from 1910, when he was technically still a student at Manchester, until 1948. Kintner was a polymath who taught virtually every science and math course at Manchester College at some point. He received his highest degree (MA in chemistry, 1914) from Ohio State University.

The most important factor the later life of Paul Flory is that Manchester College had an excellent chemistry professor, Carl Holl (1892–1961) (Figure 3.2). Holl was a 1916 graduate of Manchester College in chemistry under Kintner. He entered Ohio State University after the war in 1920 and received both an MS in chemistry in 1921 with a thesis "Action of ethylene on anhydrous ferric chloride and magnesium chloride" and a PhD in chemistry in 1923 with a thesis "The oxidation of the hexitols, mannitol, sorbital and dulcitol." He was appointed to the faculty at Manchester College in 1923, the first PhD on the science faculty. He served on the faculty until 1959! In addition to his duties as professor of chemistry, he also served as dean from 1927 to 1950. Carl Holl was known as a great teacher and was recognized with the Manufacturing Chemists Association 1958 College Chemistry Teaching Award.

Paul Flory entered Manchester College in the fall of 1927. He enthusiastically pursued a degree in science and mathematics. He joined the Science Club (Figure 3.3).

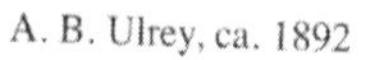
A. B. Ulrey, ca. 1892

Edward Kintner

FIGURE 3.1
Leading figures in the early history of Manchester College. (From Manchester College Archives and Brethren Historical Collection. With permission.)

FIGURE 3.2
Dean Carl Waldo Holl in 1946. (From Manchester College Archives and Brethren Historical Collection. With permission.)

FIGURE 3.3
The Manchester College Science Club for 1928. Flory at upper right in X. (From Manchester University Archives and Brethren Historical Collection. With permission.)

He chose chemistry as his major and started research with Carl Holl. By the summer of 1929, he was already looking toward his next step, but fully participated in the life of the college. Flory was a well-known name at Manchester College (Figure 3.4).

By the summer of 1930, he was ready to graduate and had become well-liked by his fellow students (Figure 3.5).

Although his picture does appear in the 1931 Yearbook, he had graduated in September 1930 and matriculated at Ohio State University. One of the organizations he was sad to leave behind was the YMCA (Figure 3.6)

His solid grounding in mathematics, chemistry, and humanity served Paul Flory (Figure 3.7) well for the rest of his life. Another serendipitous fact is that one of Flory's ('31) classmates was Roy Plunkett ('32) (Figure 3.7), the discoverer of Teflon.

When Paul Flory returned to Ohio to work at the University of Cincinnati he soon visited Manchester College. His name is a frequent entry in the home guestbook of Carl Holl. The first entry is December 13, 1939. When Flory moved to Esso Laboratories in New Jersey, he paid a visit to Holl for Thanksgiving dinner in 1941. Paul and Emily Flory attended the 25th Anniversary Celebration for Carl Holl at the graduation in May of 1948. Holl was very proud of Flory and encouraged him throughout his career. Flory's opinion of Holl was expressed as: "It is impossible for me to assess, much less put into words, his influence on me. His guidance and counsel were invaluable during my student days, to be sure, but most remarkable was his continued interest in me after I left Manchester. He maintained a personal interest in me. He had more confidence in my getting through graduate school than

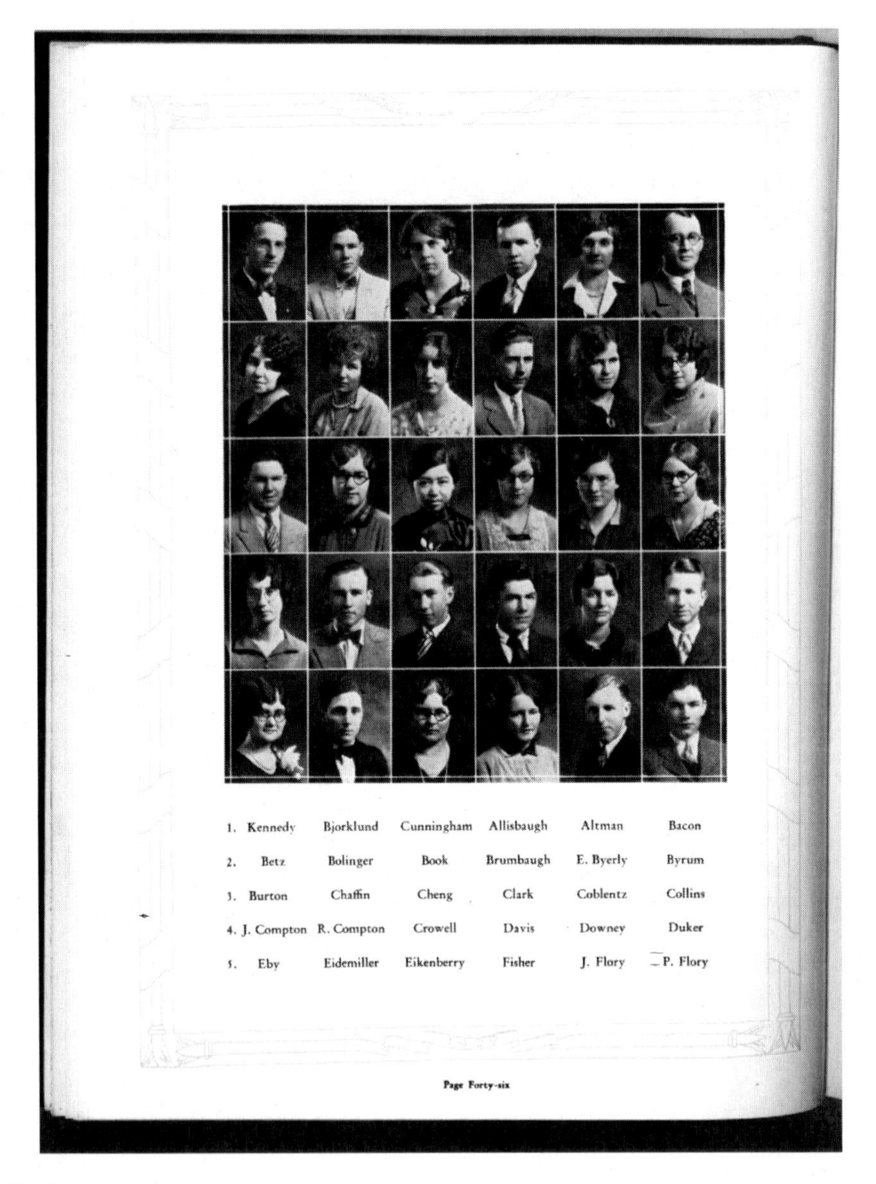

FIGURE 3.4
Sophomores from Manchester College 1929. (From Manchester University Archives and Brethren Historical Collection. With permission.)

I did. Dr. Holl had a subtle way of imparting an expectation of one that always seemed a little beyond the evidence to support that expectation. He never gushed over students but always showed genuine interest and gave encouragement. I owe him much. It would be my utmost wish that a little of that spirit and feeling has been transmitted to some of my own students."[11]

FIGURE 3.5
Juniors at Manchester College 1930 (Flory V.P.). (From Manchester University Archives and Brethren Historical Collection. With permission.)

Officers
President—Myron S. Kennedy
Vice-president—Russell Compton
Secretary—Paul Flory
Treasurer—Kieth Howard
Sponsor—Dr. R. H. Miller

FIGURE 3.6
Manchester YMCA for 1930. (From Manchester University Archives and Brethren Historical Collection. With permission.)

FIGURE 3.7
Paul Flory ('31) and Roy Plunkett ('32). (From Manchester University Archives and Brethren Historical Collection. With permission.)

After Paul Flory received the Nobel Prize in 1974, many universities granted him a Doctor of Science degree (honoris causa). But, Manchester College conferred such a degree in 1950! It was actually not too early to recognize their most famous alumnus. The next university to grant the honorary degree was The Victoria University of Manchester (England) in 1969, still well before the Nobel Prize. All the certificates for Paul Flory's honorary doctorates are on display in the Manchester College Library (Figure 3.8), along with all his medals (including the Nobel Prize Medal) and awards (such as the Perkin Medal of the Society of the Chemical Industry).

During the 1950s Manchester College realized it needed a new and modern science building in order to continue its progress in the sciences. A major campaign was initiated with a convocation in March 1955. The featured speaker was Harold Urey (1893–1981, Nobel 1934). Urey was raised in Indiana and was well aware of the role Manchester College had played in the educational history of the Hoosier State. Ground was broken in March 1958 and the building was dedicated on March 25, 1960. The featured speaker at the dedication of the Holl–Kintner Hall of Science was Paul John Flory ('31). His address entitled "The values of science" was typical of his informal speeches on science. It was written out in detail and contained well-crafted phrases. I can just hear him:

> Scientific inquiry is motivated by curiosity concerning the unknown which lies beyond; insights and imagination of the scholar are its guide. The theories of science, its abstractions and its methods are designed for comprehension of the recondite secrets of the natural order of things.[13]

FIGURE 3.8
Library display at Manchester University. (GDP photo.)

FIGURE 3.9
Cover of January 1975 Manchester College Bulletin. (From Manchester University Archives and Brethren Historical Collection. With permission.)

Manchester College

Alumni Award

Upon recommendation of the Alumni Association this Award is presented by the College to

Paul J. Flory

in recognition of significant achievements and services which reflect honor upon Manchester College.

Given at North Manchester, Indiana, on the twenty-fourth day of May, nineteen hundred and seventy-five.

President of the College

FIGURE 3.10
Manchester College alumni award for 1975. (From Manchester University Archives and Brethren Historical Collection. With permission.)

The year 1975 was a busy one for Paul Flory and Manchester College.

The January 1975 issue of the Manchester Bulletin was dedicated to Paul Flory (Figure 3.9).

Flory was chosen as the commencement speaker for that year, and was given an alumni award (Figure 3.10).

In the fall of 1975, a major celebration event to honor Paul Flory was held on October 24. Many Manchester College alumni were speakers, including Roy Plunkett ('32) and Allan Shultz ('48, Flory's graduate student at Cornell). It was a fitting finale to a great year (Figure 3.11).

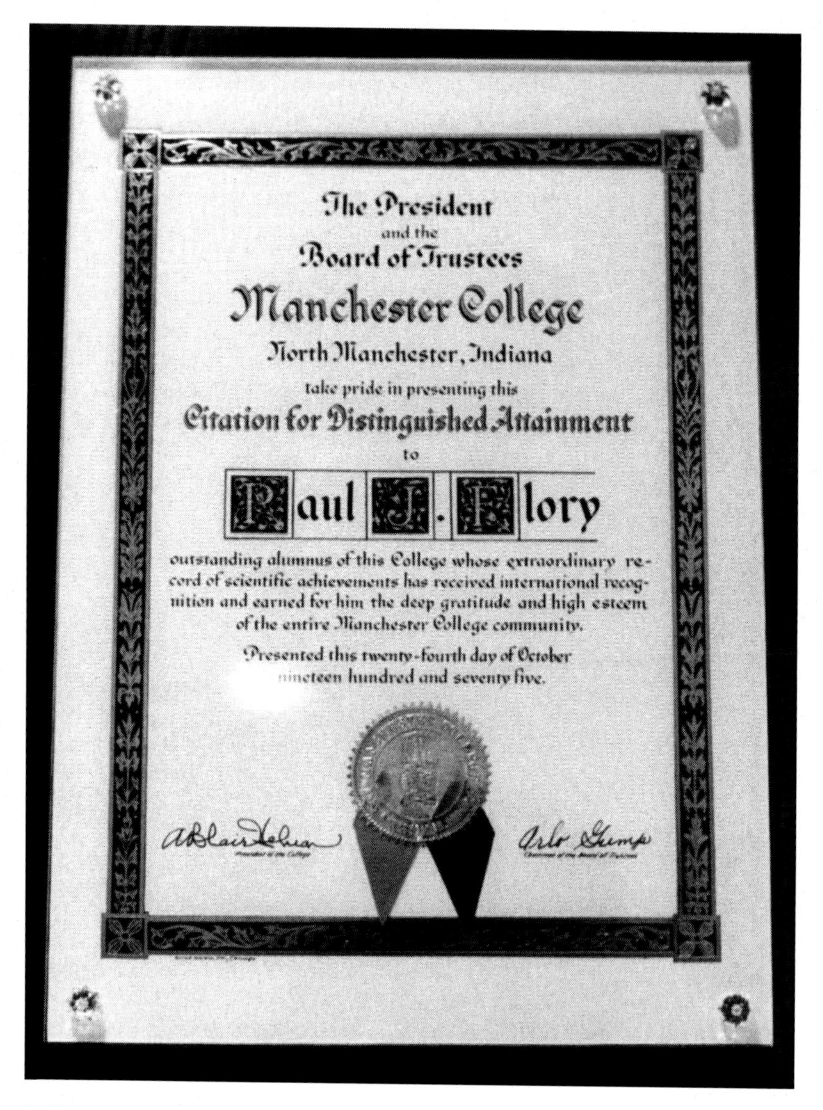

FIGURE 3.11
Citation for distinguished achievement given to Paul Flory at Manchester College on October 24, 1975. (GDP photo.)

4

Flory at Ohio State University

Always eager to move forward, Paul Flory matriculated at Ohio State University in September 1930. He applied himself diligently to his courses and obtained a master's degree in 1931. His course notebooks are available at the Chemical Heritage Foundation, and will be discussed below. The Ohio State Chemistry Department was composed of 15 faculty members in 1931 (Figure 4.1). They were distributed among the various areas of chemistry.

Introductory courses were taken in Organic Chemistry Research Laboratory and Organic Chemistry Seminar. Paul Flory notes that he prepared a "black gummy polymer." His notebooks for physical chemistry contained a different tone. His casual remarks were even lyrical when he discussed classical physical chemistry. He was fascinated by the classic thermodynamics text by Lewis and Randall.[14]

The second year was devoted to more advanced topics in physical chemistry. Edward Mack taught an advanced course in phenomenological physical chemistry based on the classic text by Adam.[15] This thorough discussion of the observable properties of actual chemical systems provided a world of chemistry for his entire career. The winter quarter was devoted to chemical kinetics. The text was not a watered down book intended for undergraduates. The monograph by Hinshelwood was assigned.[16] In order to become a full physical chemist, Paul Flory took a series of advanced courses in the physics department. Quantum mechanics was taught by Alfred Lande.

Finally in 1932 Paul Flory took several courses with Herrick Johnston (Figure 4.2). A historical and technical course on atomic structure was followed by advanced courses on thermodynamics and molecular structure. Johnston agreed to accept Paul Flory as a PhD student.

The 1930s was a great age for chemical kinetics. Paul Flory studied the photochemistry of nitric oxide. He defended his thesis, "The photo-decomposition of nitric oxide," in 1934 (Figure 4.3).

Just as with Manchester College, Paul Flory made an impression on the chemistry faculty. When he had progressed in his career to the point where he was the department head of Fundamental Polymer Research at the Goodyear Tire and Rubber Company in Akron, Ohio, he was awarded the Sullivant Medal in 1945. This prestigious award was given every 5 years to an Ohio State graduate who had made a major contribution to American life.

FIGURE 4.1
Ohio State University Chemistry Department in 1931. (From Ohio State Archives. With permission.) (Back row [l–r]): Melville Wolfram, G. Bryant Bachman, Marion Hollingsworth, W. Conrad Fernelius, Herrick L. Johnson, Harvey Moyer, Wallace Brode. (Front row [l–r]): Edward Mack, Jr., Cecil Boord, William McPherson, William Evans, William Henderson, C.W. Foulk, Jesse Day, Wesley France.

FIGURE 4.2
Herrick L. Johnston (1898–1965). (From Ohio State Archives. With permission.)

The Ohio State University
hereby confers upon
Paul John Flory, B.S., M.Sc.
the degree of
Doctor of Philosophy
together with all the rights, privileges and honors appertaining thereto in consideration of the satisfactory completion of the course prescribed in
The Graduate School
In Testimony Whereof, the seal of the University and the signatures as authorized by the Board of Trustees are hereunto affixed.
Given at Columbus on the eleventh day of June in the year of our Lord nineteen hundred thirty-four

FIGURE 4.3
Diploma from The Ohio State University (1934). (GDP photo.)

FIGURE 4.4

Bronze plaque commemorating Paul J. Flory at Ohio State University. (GDP photo.)

For example, Arthur Schlesinger of Harvard won this award. Ohio State remembered Paul Flory again in 1970 with an honorary Doctorate of Science. Lest anyone forget that Paul Flory graduated from Ohio State University, a bronze plaque was placed outside the McPherson Chemistry Building in 1985 (Figure 4.4).

5

Flory at DuPont 1934–1938

E.I. du Pont de Nemours & Company is one of the oldest U.S. corporations. It has its roots in the French laboratory (and bedroom) of Antoine Lavoisier. The initial manufacturing efforts were focused on gun powder and dynamite. But the biggest bang occurred when the research and development laboratory was founded in 1902.

"In 1927, Chemical Director Charles M.A. Stine (Figure 5.1) began an experiment in pure science or fundamental research."[17] One of the fields chosen for exploitation was polymers, and Wallace A. Carothers (Figure 5.2) was lured away from Harvard in 1928 to spearhead this effort. Carothers flourished in this environment and by the time he gave the plenary lecture at the Faraday Discussion on Polymerization in 1935 the field of polymer science was firmly established.[18]

The vision of Charles Stine led to the construction of "Purity Hall" and enough money to hire 25 scientists! It was very difficult to lure truly academic scientists away from the university womb, but Wallace Carothers (1896–1937) had already been seduced by the writings of Robert Kennedy Duncan (1868–1914) of the Mellon Institute into believing that great results would come from combining the best of fundamental research with an industrial environment. The freedom to pursue the advancement of knowledge in areas of industrial interest satisfied both groups. It was this exciting opportunity that drew Paul J. Flory to commit his early career to Carothers' polymer department.

Unfortunately, very soon after Flory's arrival in 1934, Carothers suffered a breakdown and was never truly able to converse effectively again. After Carothers committed suicide in 1937, Flory was looking for a new home. The collapse of the entire polymerization department left little opportunity for a rapidly maturing polymer theorist! Since Carothers was not effective during this period, Flory was often called on to write the internal technical reports. He learned a great deal about industrial life and the interplay between chemical knowledge and profitable products.

Nevertheless, Carothers did inspire Flory to initiate a program in polymer kinetics. Flory had done his doctoral work on chemical kinetics. He mastered the literature of polymer kinetics, as it existed in 1934, and developed mathematical methods to calculate the distribution of chain lengths for

FIGURE 5.1
Charles M.A. Stine (1882–1954). (From DuPont. With permission.)

condensation polymers. Carothers made mention of this work in his magisterial lecture of 1935, mentioned above. It appeared in print as

> "Molecular Size Distribution in Linear Condensation Polymers," *J. Am. Chem. Soc.*, 58, 1877, 1936.

Polymers formed by radical polymerization are observed to have a very broad distribution of molecular weight. Many workers had tried to

FIGURE 5.2
Wallace H. Carothers (1896–1937).

understand the reason for this phenomenon, but it was Paul Flory that figured it out. Growing radical chains can terminate by many processes. Flory identified chain transfer to monomer or other chains as the reason that the polydispersity of the chains was so high. This process produced another growing chain, rather than terminating two chains by coupling or disproportionation. The classic paper on "The mechanism of vinyl polymerization" appeared in *JACS* in 1937.[19]

Another factor that led Flory to look for greener pastures was the change in the research atmosphere at the Experimental Station at DuPont.

"When Bolton clamped down on publication at the Station in the late 1930s and early 1940s, he also scotched presentation of papers because this too constituted disclosure."[17] Flory was driven by an unfettered search for scientific truth, and was less disposed to confine his thoughts, writings, and conversations to legally sanitized channels. Nevertheless, his departure was amicable and when he again occupied an academic professorship at Cornell, Flory was retained as a consultant until his death in 1985.

6

Flory at the University of Cincinnati 1938–1940

CONTENTS

Transition from DuPont

Paul Flory was strongly affected by Carothers' suicide on April 29, 1937.[20] In an interview,[21] Flory referred to Carothers' death as "one of the most profoundly shocking events in my life … It just pulled the rug from under my hopes … I realized how much a shield he had been and much of an influence … when he was gone." After Carothers' death, it was clear to Flory that he had lost an important mentor and felt that there was now no one devoted to fundamental research at DuPont. In 1938, Flory decided to leave DuPont and joined the Basic Sciences Research Laboratory of the University of Cincinnati (UC) as a research associate. The Basic Sciences Laboratory was the perfect opportunity for Flory at that time. Flory again had the freedom to develop his own ideas, and from 1938 until 1940, he continued his important studies of polymer viscosity and the mechanism of chain-length distributions of polymers that he had begun at DuPont. A particularly important milestone was the development of a mathematical model for gelation.

The UC and the Basic Sciences Research Laboratory

The University of Cincinnati was founded in 1819 as Cincinnati College, about the same time as the Medical College of Ohio was established.[22] By the

FIGURE 6.1
(Left) photo of Herman Schneider (*co-op education*); (right) bust of Herman Schneider in front of the original engineering building, Baldwin Hall, on the UC campus today. (Photograph is the courtesy of Katie Hageman, University of Cincinnati.)

early 1900s, Cincinnati College, the Medical College, the Cincinnati College of Law, and the Queen City College of Pharmacy became part of what is now UC. A Teachers College was added in 1905 and a Graduate School was established in the College of Arts and Sciences in 1906. In that same year, Herman Schneider (Figure 6.1), a civil engineering professor, established the first cooperative education program that remains one of the largest and most successful co-op programs in the United States. Schneider's plan was to combine university instruction with practical applications in the working world.

The Institute of Scientific Research expanded the concept of co-op education by joining the university with industry on a campus setting. By 1924, the institute consisted of three laboratories—the Leather Research Laboratory (1920) of the Tanners' Council of America, the Research Laboratory of the Lithographic Technical Foundation (1924), and the Basic Science Research Laboratory (1924). The Tanners' Laboratory investigated problems in chemical science for the benefit of the tanning industry and still exists today (http://www.leatherusa.org/). The objective of the Lithographic Laboratory was to explore the scientific principles of lithography. The Basic Science Research Laboratory differed from the Tanners' and Lithographic laboratories in its freedom to investigate any research problem that the staff considered to be promising. Researchers at the laboratory came from a variety of scientific disciplines to generate creative and useful research through interdisciplinary collaboration. The Basic Science Research Laboratory was originally directed by Dr. George Sperti, a 1923 graduate of UC in electrical engineering. Among his contributions was the sun lamp and frozen orange juice concentrate. Sperti left UC in 1935, a little over 2 years before Flory arrived, to join Cincinnati Archbishop John McNicholas to establish the Institutum Divi Thomae, a graduate school where scientific research was conducted. His work there led to the development of pain reliever Preparation H® and Aspercreme®. When George Sperti left the Basic Science Laboratory in 1935, Dr. Walter Soller became the director of its graduate programs, sponsored primarily by the Stephen Wilder Endowment Fund, in which students spent alternative 6-month periods in the laboratory and at industrial research locations. The laboratory moved from the attic of Cunningham Hall to a wing in Hanna Hall on the UC campus, and then to its own building behind McMicken Hall that still stands today. It was renamed the Applied Science Research Laboratory as an independent unit of the College of Engineering. So successful was this program that the UC board of directors authorized a Graduate Department of Applied Science in 1940 about the time that Flory left for a position as a research chemist at the Esso (later Exxon) Laboratories of the Standard Oil Development Company in Linden, New Jersey. The new department acquired the functions of the Basic Science Laboratory and offered a Master of Science in Applied Science and a Doctor of Science, with the opportunity for co-op experience. The Lithographic Research Laboratory and the Tanners' Laboratory then became independent units as well. It is in view of this unique opportunity to pursue basic research in a setting of interdisciplinary collaboration and industrial partnership that Flory found a new home from 1938 to 1940 to continue the work he started at DuPont.

Scientific Contributions

At UC, Flory focused on three important areas in the early development of polymer science:

1. A continuation of theoretical and experimental studies of melt viscosity begun at DuPont and an extension to solutions.
2. Experimental verification of the important *equal reactivity principle* that states the reactivity of functional groups in polycondensation reactions are dependent only on the chemical environment of each functional group but are independent of the size of the polymer chain.
3. The initial development of a mathematical model of *gelation*, whereby monomers containing more than two functional groups can polymerize into a highly branched structure and form a gel at a specific stage (i.e., the gel point) in the polymerization process.

TABLE 6.1

Publications by Flory Based upon Research Conducted While at the University of Cincinnati

Author(s)	Article Title	Journal
P. J. Flory	Intramolecular reaction between neighboring substituents of vinyl polymers[a]	*JACS* 61, 1518–1521 (1939)[23]
P. J. Flory	Kinetics of polyesterification: A study of the effects of molecular weight and viscosity on reaction rate[a]	*JACS* 61, 3334–3340 (1939)[24]
P. J. Flory	Viscosities of linear polyesters. An exact relationship between viscosity and chain length[a]	*JACS* 62, 1057–1070 (1940)[25]
P. J. Flory	Molecular size distribution in ethylene oxide polymers[a]	*JACS* 62, 1561–1565 (1940)[26]
P. J. Flory	Kinetics of the degradation of polyesters by alcohols[a]	*JACS* 62, 2255–2261 (1940)[27]
P. J. Flory	A comparison of esterification and ester interchange kinetics[a]	*JACS* 62, 2261–2264 (1940)[28]
P. J. Flory and P. B. Stickney	Viscosities of polyester solutions and the Staudinger equation[a]	*JACS* 62, 3032–3038 (1940)[29]
P. J. Flory	Molecular size distribution in three-dimensional polymers. I. Gelation[b]	*JACS* 63, 3083–3090 (1941)[30]

[a] Contribution from the Basic Sciences Research Laboratory, University of Cincinnati.
[b] Contribution from the Graduate School of Applied Science of the University of Cincinnati.

This work was published between 1939 and 1941 in a series of eight articles published in the *Journal of the American Chemical Society* (JACS) as shown in Table 6.1. A chronological review of these contributions is given in the sections below.

Statistics of Intramolecular Reactivity in Vinyl Polymers

In his first paper issued from the Basic Sciences Research Laboratory in 1939,[23] Flory focused on the reactivity of functional groups (R) of vinyl polymers having the repeat unit

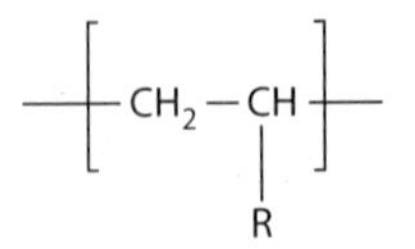

and calculated the probability of *intramolecular* reactions for head-to-tail, head-to-head, and tail-to-tail structures.* The example given by Flory was a high-temperature reaction of poly(methyl vinyl ketone) in its head-to-tail arrangement where an intramolecular condensation of its substituent groups

* Head-to-tail arrangement is the most probable case.

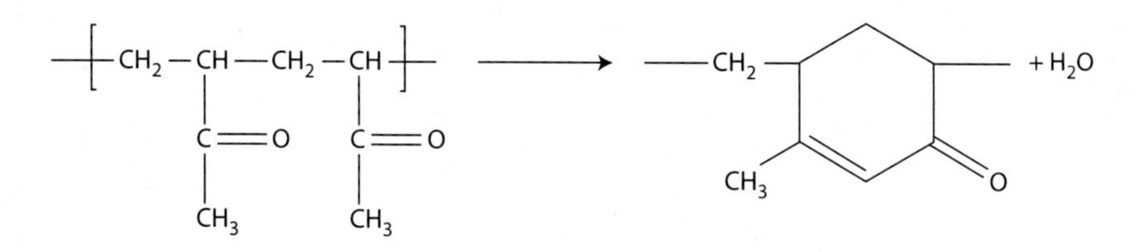

FIGURE 6.2
Illustration of an intramolecular reaction between neighboring substituent groups.[23] Heating of poly(methyl vinyl ketone) at about 300°C causes the carbonyl group of one repeat unit to condense with the methyl group of the contiguous substituent to form a cyclic structure in the polymer backbone.

(–$COCH_3$) can occur as illustrated in Figure 6.2. In this example, where, Flory showed that 13.33% of the substituent groups are prevented from reacting due to isolation between reacted pairs in agreement with experimental values.

Kinetics of Polyesterification and Degradation

In his second 1939 publication,[24] Flory reported measurements of the kinetics of polyesterification reactions including those of diethylene glycol and adipic acid, decamethylene glycol and adipic acid, and preliminary studies of diethylene glycol and adipic acid. Reaction kinetics of the polyesterifications was compared with nonpolymer forming esterifications. Conclusions of these studies were that the reaction rate of polyesterification reactions was not affected by either the increase in molecular weight or by the concurrent increase in viscosity. The slow rate of the polyesterification when molecular weight was large was attributed to the third-order character of the esterification reactions. These results supported Flory's earlier postulate, the *equal reactivity principle,* that the reactivity of functional groups were independent of molecular weight.[31]

Polymer Viscosity and Polyesters

While at UC, Flory also focused his attention on melt and later solution viscosity and published several important journal articles during 1940. In the first of these publications,[25] Flory showed for a series of low-molecular-weight polyesters (molecular weights between 200 and 10,000) that the melt viscosity depended upon the weight-average molecular weight ($\bar{M}_w$) through the relationship

$$\log \eta = A + C'\bar{M}_w^{1/2} \quad (6.1)$$

where A and C' are constants. Alternatively, this relationship can be expressed as a function of the weight-average chain length, $\bar{Z}_w$, as

$$\log \eta = A + C\bar{Z}_w^{1/2} \quad (6.2)$$

In a subsequent publication of 1940, Flory, in collaboration with P. B. Stickney,* investigated the viscosity of polyester *solutions* in relation to the expression proposed by Nobel Laureate Hermann Staudinger† between specific viscosity (η_{sp})‡ and molecular weight (M)

$$\eta_{sp}/c = K_m M \tag{6.3}$$

where c is concentration (units of mol/L) and K_m is a constant. In their study of the viscosity of dilute solutions of 12 decamethylene adipate polyesters (weight-average molecular weight between 1500 and 30,000) in diethyl succinate and chlorobenzene,[29] Flory and Stickney demonstrated that the proper relationship between the relative viscosity,§ η_r, and $\bar{M}_w$ had the form

$$[(\ln \eta_r)/c]_o = K_w \bar{M}_w + I \tag{6.4}$$

K_w and I are constants and demonstrated that the relationship given by Staudinger, Equation 6.3, did not fit the experimental data.

In another study, Flory reported on the kinetics of *degradation* of polyesters by alcohols,[27] using the example of the degradation of decamethylene adipate polyesters by decamethylene glycol and by lauryl alcohol as generically illustrated below.

$$\ldots ORO{-}\overset{O}{\overset{\|}{C}}R'\overset{O}{\overset{\|}{C}}{-}ORO\ldots + R''{-}OH \longrightarrow \ldots OROH + R''O{-}\overset{O}{\overset{\|}{C}}R'\overset{O}{\overset{\|}{C}}{-}ORO\ldots$$

Flory showed that the kinetics of the degradation process can be followed by viscosity measurements using Equation 6.2. In an accompanying paper, Flory showed that viscosity measurements could also be used to compare the kinetics of esterification and ester interchange.[28]

Molecular-Weight Distribution

Also in 1940,[26] Flory reported studies on the molecular-weight distribution in ethylene oxide polymers formed by the reaction of ethylene oxide with ethylene glycol as illustrated below.

* Work was published when Flory had left for Esso Laboratories and Stickney was at Batelle Memorial Institute in Columbus Ohio.

† Recipient of the 1953 Nobel Prize in Chemistry.

‡ The specific viscosity is defined as

$$\eta_{sp} = \frac{\eta - \eta_s}{\eta_s}$$

where η is the viscosity of the dilute solution and η_s is the viscosity of the pure solvent.

§ $\eta_r = \eta/\eta_s$.

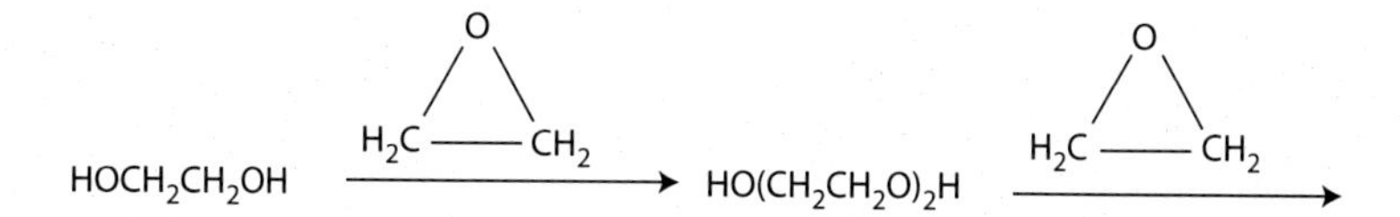

By statistical arguments, Flory showed that in this case, where polymer formation occurs by the addition of monomer to a fixed number of polymer molecules, the number of different molecular-weight species followed a *Poisson distribution* for the chain length given as

$$\frac{N_x}{N^o} = \frac{\exp(-\nu)\nu^{x-1}}{(x-1)!} \tag{6.5}$$

Using Flory's nomenclature, N_x represents the number of x-mers, N^o is the number of propagating molecules, and ν is the ratio of monomer molecules consumed at a time t to N^o in the form

$$\nu = \frac{\Delta m}{N^o} \tag{6.6}$$

Gelation

An example of a typical polycondensation reaction is the polycondensation of dimethyl terephthalate with ethylene glycol to form poly(ethylene terephthalate) (PET) as illustrated in Figure 6.3a. In this example, the two starting monomers have difunctional reactive groups. While at UC, Flory started to redirect his attention to polycondensation reactions involving monomers that had three or more functional groups that could be used to form a branched network. These "three-dimensional" polymers, form an insoluble, elastic network having a gel-like consistency. This process was termed *gelation*. The point of gelation occurs well before all monomers are consumed in the polycondensation reaction. Important examples of three-dimensional polymers include glycerol polyesters such as glyptal,* made from the polycondensation of glycerol (a trifunctional monomer having three reactive hydroxyl groups) and phthalic anhydride as illustrated in Figure 6.3b. The last publication from Flory's work at UC (submitted while at Esso)[30] was the first of several publications on gelation including one published shortly afterwards from Esso.[32]

Flory proposed that gelation is mathematically similar to the kinetics of explosions and nuclear chain reactions. Flory used this analogy to develop a mathematical model that showed that gelation occurred at a specific gel point. Flory's model, and later the work of W. H. Stockmayer[33] based on it,

* Developed as a protective coating resin by GE in 1912.

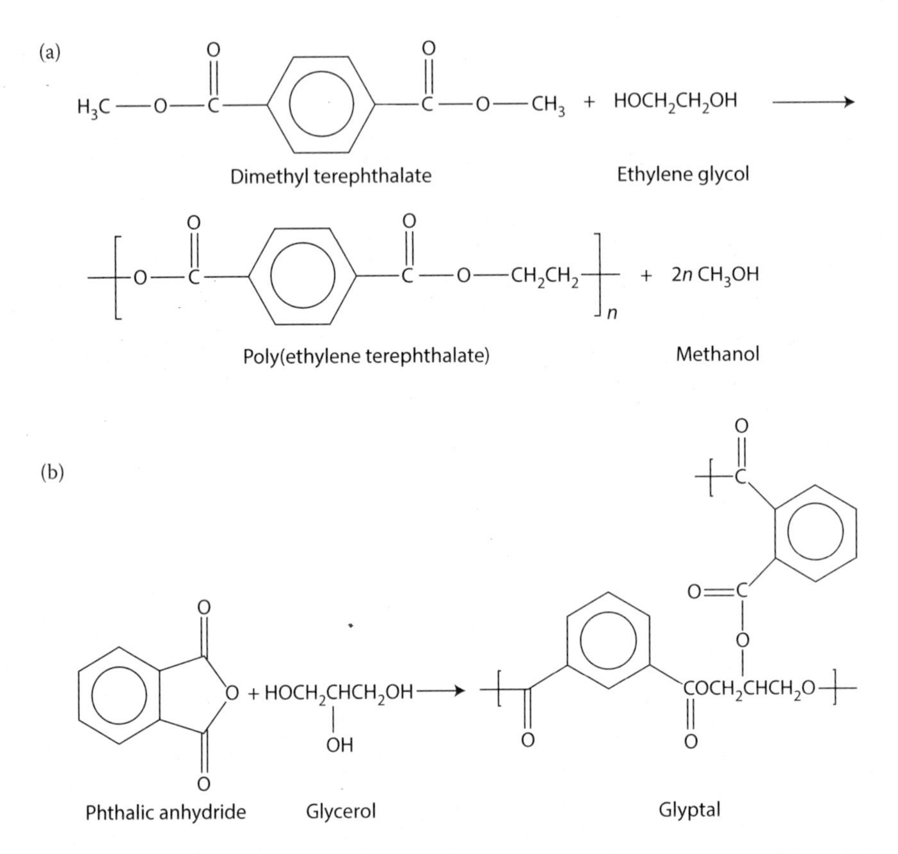

FIGURE 6.3
(a) Polycondensation of poly(ethylene terephthalate) and (b) polycondensation of glyptal resin.

provided a quantitative description of gelation. This development found application not only in the field of synthetic polymers but also in biological systems such as antigen–antibody reactions and to the chemistry of inorganic hydrous systems. Although Carothers had correctly concluded that gelation corresponded to the formation of a network of essentially infinite molecular weight,[34] his calculations using simple stoichiometry arguments suggested that the number-average molecular weight was the appropriate signal for gelation. In fact, the gel point is found to occur much earlier, when the number-average molecular weight is still modest. Here, Flory recognized that the branched polymers would have a size distribution much broader than that of linear polymers and that the gel point corresponds to a diverging weight-average molecular weight. As reported in a series of three papers published in 1941,[30,35] Flory presented his quantitative theory of the gel point and of the distribution of molecular weights. In a 1946 review of condensation polymerization, Flory has provided a later perspective of his work and that of Carothers in the development of the theory of gelation.[36]

FIGURE 6.4
Flory's lecture at UC in 1977. (From the University of Cincinnati. With permission.)

Later Association with UC

After Flory received the Nobel Prize for Chemistry in 1974, Flory made several return visits to Cincinnati. One of these visits was in 1977 when Flory gave a seminar (Figure 6.4) and met with members of the newly formed Polymer Research Center (PRC) established by James Mark after he left the University of Michigan in that year (Figure 6.5).

Honorary Doctorate

Flory received an honorary doctorate at UC on June 13, 1982. Neil Armstrong also received the honorary degree of Doctor of Science and was speaker at UC's commencement ceremony in front of an audience of more than 20,000. Armstrong had been an adjunct professor of aerospace engineering from 1971 to 1980. UC's President, Henry Winkler, awarded the honorary degree to Flory giving the following oration:

> The fruits of your intelligence, perseverance and ingenuity surround us, in the polymeric substances utilized today in clothing, paint, films, toys, and machine parts, yet until you found the key, the outstanding potential of these substances was unfulfilled, Through your investigation, in academe and industry alike, a great wealth of synthetic materials has been brought into the service of mankind.

FIGURE 6.5
Meeting of Flory with the executive board of the PRC at the University of Cincinnati in 1977. (from left to right) Flory (seated), James E. Mark (director of the PRC), Richard P. Chartoff, Ryong-Joon Roe, and James F. Boerio. After Richard Chartoff left UC for the University of Dayton, Joel Fried joined the executive board of the PRC in December of 1978 from Monsanto CR&D (St. Louis) and later became the second director of the PRC from 1989 to 1992. (Reproduced from a *UC This Week* article published in November 22, 1977. Photograph courtesy of the Archives and Rare Books Library, University of Cincinnati.)

Multitudinous honors have been bestowed upon you. Rightfully so, and our pride as a university grows with each reward, remembering that the laboratories on this campus were the site of your early research.

Today, in laboratories throughout the world, the physical methods which you developed, continue to be applied in the search for novel and commodious materials. Among these laboratories is the University of Cincinnati Polymer Research Center, toward which you have contributed no small measure of assistance, an example of your commitment to furthering the opportunities for others to expand upon the firm foundation of your own discoveries.

In recognition of the unquestionable importance of your accomplishments, and by virtue of the authority vested in me as the President of the University of Cincinnati, it is my honor to confer upon you, Paul J. Flory, the degree of Doctor of Science, honoris causa.

Henry R. Winkler
President

7

Flory at Esso 1940–1943

Although Paul Flory preferred an unfettered research environment, he could not refuse a good industrial offer from the Esso Laboratories (Figure 7.1) of the Standard Oil Development Company in Elizabeth, New Jersey. He joined them in 1940.

One of the areas of polymer science pursued by Paul Flory while at Esso was the thermodynamics of two component liquid mixtures. In 1940, this area of research was still in its infancy. A series of papers published in the *Journal of Chemical Physics* changed all that. The first major contribution (1941) was the derivation of an expression for the entropy of mixing that involved the volume fraction rather than the mole fraction of the solvent.[37] Standard physical chemistry texts still fail to acknowledge this feature, but industrial laboratories, where empirical adequacy is still an important issue, use the Flory expression. A much more extensive paper that included the full discussion of the so-called Flory–Huggins theory appeared in 1942.[38] When Paul Flory knew that Maurice Huggins (1897–1981) (Figure 7.2), then at Kodak Laboratories, was working in this area, he approached him and suggested collaboration. Huggins demurred and encouraged Flory to publish separately. Flory and Huggins remained good friends throughout their careers, and Maurice was a welcome visitor in the Flory laboratory at Stanford when he retired to Menlo Park, California in the 1960s. Huggins had been a professor at Stanford from 1926 to 1933 and received his degrees in chemistry from the University of California at Berkeley (BS 1919, PhD 1922).

The theory assumes a mean-field distribution of segments in the solution. This is acceptable for a mixture of small molecules, but for a large chain molecule the polymer concentration inside the coil is much larger than the average for the solution as a whole. A more detailed discussion of this issue occurs in the chapter on Flory at Cornell. The Flory–Huggins theory allows a calculation of the phase diagram for a two-component liquid mixture. It is qualitatively correct, like the van der Waals theory of the liquid–gas phase transition, but since it ignores the volume change on mixing and assumes random mixing, it is not quantitative. It also cannot predict the branch at high temperature due to the large negative volume change on mixing a small molecule and a polymer near the critical point of the solvent. This issue is addressed in the chapter on Flory at Stanford.

One of the most pressing problems in polymer science in the early 1940s was the quantitative theory of rubber elasticity. Substantial conceptual

FIGURE 7.1
Esso research in the early 1940s. (Adapted from Chemical Heritage Foundation.)

progress had been made by Werner Kuhn (1899–1963) and by Eugene Guth (1905–1990) and Herman Mark (1895–1992), but a readily applicable expression was needed to guide further work. Flory collaborated with John Rehner, Jr. (1908–1997) (Figure 7.4) on a classic series of papers on polymer networks and gels. Flory considered in detail the role of the network in the theory. He assumed that the distribution of end-to-end vectors for the chains between crosslinks was modified by the affine deformation and detailed equations were derived.[39] The theory of rubber elasticity is still not finished today, but this work of Flory was a good step in the right direction.

With both a workable theory for network deformation, and a semiquantitative theory of polymer solutions, Flory and Rehner[40] tackled the problem of swollen rubber. Their combined theory is so good, that it is still used in

FIGURE 7.2
Maurice Loyal Huggins (1897–1981). (From Interscience. With permission.)

industry. They even correctly described the change in the elongational modulus due to swelling (Figure 7.3).

John Rehner, Jr. was born at home in New York City on October 29, 1908. His family moved to Kansas City and he attended the University of Missouri (BS chemical engineering, 1929; MA chemistry, 1930). He obtained a PhD in physical chemistry from the University of Minnesota in 1933, with a minor

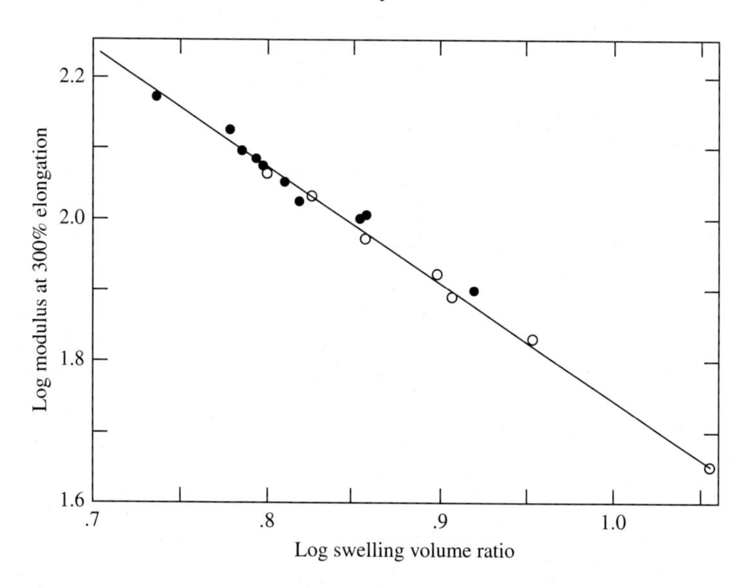

FIGURE 7.3
Elongational modulus as a function of swelling. (Reprinted with permission from Flory, P.J., Rehner, J. Jr., Statistical mechanics of cross-linked polymer networks: II, swelling, *J. Chem. Phys.*, 11, 521. Copyright 1943, American Institute of Physics.)

FIGURE 7.4
John Rehner, Jr. (From Roberta Rehner Iverson. With permission.)

in physics. Rehner's interest in polymer physics led him to be a founding member of the Division of High Polymer Physics (DHPP) of the American Physical Society in 1943. He went on to chair DHPP in 1960–1961.

His first professional position was with the B.F. Goodrich Company. He served in both the Akron and Kuala Lumpur laboratories! He knew rubber "from the ground up." He returned to the mainland at the Esso Research and Engineering Company, Linden, New Jersey in 1942. He retired from there in 1969, after moving up through the ranks to senior research associate (Figure 7.4).

8

Flory at Goodyear Tire and Rubber Company 1943–1948

Although the name of Charles Goodyear (1800–1860) is synonymous with vulcanized rubber,[41] the current company that bears his name has no relation to the historical character. The present company was founded in 1898 by Frank A. Seiberling (1859–1955). Another leading figure in the history of Goodyear was Paul W. Litchfield (1875–1959), an 1896 graduate of the Massachusetts Institute of Technology. The business was a success and in 1916 became the world's largest tire company. Litchfield's tenure as president (1926–1940) brought many innovations.

The 1940s saw the presidency of Edwin J. Thomas (1899–1987). Goodyear was fully committed to the war effort, even before the United States entered the war. As part of its commitment to the Rubber Reserve Program,[42] Goodyear constructed a new Research Laboratory in Akron in 1943. One of their first new hires for this facility was Paul J. Flory. The director of research during this period was Lorin B. Sebrell (1894–1984). He was a fixture in the Rubber Division of the American Chemical Society, and served as chairman in 1933. He was awarded the Charles Goodyear Medal in 1942, the second such recipient. Paul Flory won this medal in 1968. Two other members of the management team were Harold J. Osterhof (1897–1982) and James D'Ianni (1914–2007), both Goodyear medalists and future directors of research at Goodyear.

Each company involved with the Rubber Reserve was responsible to send up to four people to the Rubber (later polymer) Research Discussion Group every 3 months. Goodyear chose Flory as one of their representatives and appointed him as Head of the Fundamental Polymer Research Department. One of the benefits for Flory was that he was in regular contact with the brightest and best scientists in America: Carl Marvel and Fred Wall (Illinois), Izaak Kolthoff (Minnesota), W.D. Harkins and Morris Kharasch (Chicago), William O. Baker (Bell Laboratories), and Peter Debye (Cornell).

Paul Flory wasted no time in advancing his research agenda at Goodyear. He was immediately invited to give the plenary lecture in the Symposium on the "Theory of long-range elasticity" at the Cleveland ACS Meeting in 1944.[43] It summarized all the work done through 1944 and presented new comparisons between theory and experiment. He was also tapped to produce a critical review article on "Condensation polymerization."[44] He established a true exemplar for an industrial pure scientist: He was expected to know everything about his field of expertise and to be advancing it at the highest level.

Flory was keenly aware that the Flory–Huggins theory had serious shortcomings for the description of very dilute polymer solutions. The assumption of a uniform distribution of polymer segments was greatly in error when the amount of volume pervaded by the polymer coils was small in comparison to the total volume. Inside a polymer coil, the concentration was equal to the limiting value for a random coil; it was zero elsewhere. Flory published the first paper outlining better strategies for calculating the chemical potential for a polymer molecule in dilute solution.[45]

At the other end of the concentration range, Flory considered the phenomenon of strain-induced crystallization in natural rubber. The key concept was the change in the entropy of the amorphous rubber phase due to the alignment of the chains produced by stretching. The change in the melting temperature can then be calculated from the Clapeyron equation. One of the differences between the crystallization of a pure liquid and a cross-linked rubber is that the rubber does not fully crystallize. The cross-links are excluded from the crystal and serve as an "impurity."[46]

One of Paul Flory's functions at Goodyear was to think about long-term and large issues. He issued an important memorandum to Lorin Sebrell in April 1944 on "Hydrocarbon polymers for non-rubber uses."[47]

The focus of this article is on high-pressure polymethylene. It is an excellent insulator with a low dielectric loss. It can also be blended with polyisobutylene to produce a more flexible product. Flory's intimate knowledge of actual polyisobutylene was demonstrated for me. He could determine the approximate molecular weight just by touching the bulk sample and feeling its tackiness. Progress in preparing better polymers of polymethylene was discussed and catalytic approaches were suggested. Flory never forgot his background in chemical kinetics and no reaction was beyond his consideration.

FIGURE 8.1
Norman Rabjohn (1915–2000). (From Goodyear Archives. With permission.)

FIGURE 8.2
John R. Schaefgen. (From the Chemical Heritage Foundation. With permission.)

Goodyear was also a place where Paul Flory met many scientific friends in his research group. One of them was the synthetic chemist Norman Rabjohn (1915–2000) (Figure 8.1).

Another friend was John R. Schaefgen (1918–). In his oral history at the Chemical Heritage Foundation he commented about Flory: "My teacher, my leader, my idol. We thought a lot of him. We thought at that time that he would be a Nobel Prize winner."[48] Eventually, Schaefgen went on to a distinguished career at DuPont (Figure 8.2).

The most important friend and collaborator that Paul Flory met at Goodyear was Thomas G Fox (1921–1977). They published many papers together. Much more will be said about this friendship in later chapters. Fundamental studies of the temperature and molecular weight dependence of the pure liquid viscosity of polymers were carried out.

After the war was concluded, the research atmosphere at Goodyear reverted to more traditional antipathy toward truly fundamental work. Paul Flory is infamous for commenting, when asked why he left Goodyear, "I grew tired of casting synthetic pearls before real swine!"[49]

9

Flory at Cornell University 1948–1956

Cornell University owes its existence to the fortune of New York State Senator Ezra Cornell and the academic vision of Senator Andrew Dickson White (Figure 9.1). It opened its doors to students in 1867 as the New York State Land Grant College. A.D. White was its first president. He envisioned a school where a scientific and technical education would be married with studies in history and literature.

Cornell also benefitted from the philanthropy of George Fisher Baker. When a new chemistry building was needed, Baker provided the money and laid the cornerstone in 1921.[50] When the building was finished and furnished, there was still money left over. The chemistry department instituted the George Fisher Baker Nonresident Lectureship in 1928. The list of people who have held this post includes many Nobel Prize winners, often well before the award. In 1939 the Baker Lecturer was Peter Joseph William Debye (1884–1966) (Figure 9.2). He had won the Nobel Prize in chemistry in 1936 and was delighted to leave Berlin and move to Cornell, at least officially on a leave of absence. He was the most famous chemist in Germany and was the director of the Kaiser Wilhelm Institute for Physics.

Peter Debye became the chairman of the chemistry department and set about improving the faculty. One of the brightest young faculty members present at Cornell during the Debye era was John Gamble Kirkwood (1907–1959). After receiving his PhD in chemistry from the Massachusetts Institute of Technology in 1929, he worked with Peter Debye in Leipzig during 1931–1932. Four papers were produced that year and Kirkwood's interest in electrolyte solutions never waned. He joined Cornell in 1934 as an assistant professor, left for the University of Chicago briefly and returned in 1938 as Todd Professor of Chemistry (1938–1947). Although Kirkwood had left Cornell by the time Flory arrived, the work he carried out there on the theory of transport processes was the basis of Flory's treatment of the viscosity of polymer solutions. Interestingly, Kirkwood was elected to the National Academy of Sciences in 1942, 5 years before Debye (1947) and 11 years before Flory (1953).

In 1948, Peter Debye invited Paul Flory to be the Baker Lecturer, and then encouraged him to stay. The academic position at Cornell allowed Flory to attract outstanding graduate students and postdoctoral fellows. His first such fellow was William Krigbaum (1922–1991) (Figure 9.9). They attacked the problem of the osmotic pressure of a dilute polymer solution using the statistical mechanical methods of Gibbs and Kirkwood. In order to carry out such a calculation, it was necessary to derive a potential of mean force

FIGURE 9.1
Andrew Dickson White (1832–1918), first president of Cornell University. (From Cornell University. With permission.)

between two individual polymer coils in solution. The Flory–Krigbaum potential is not quite right, but it allowed them to carry through the calculation of the second osmotic virial coefficient. It was a major step forward on this problem.

Paul Flory's first graduate student at Cornell was Allan R. Shultz (1926–2003) (Figure 9.3), a 1948 graduate of Manchester College. He explored phase

FIGURE 9.2
Peter J. W. Debye (1884–1966). (From National Academy of Sciences. With permission.)

FIGURE 9.3
Allan R. Shultz (1926–2003). (From Manchester University Archives. With permission.)

equilibria in polymer-solvent systems.[51] He went on to an industrial career with 3M and with the General Electric Company. He and his wife, Wylan, remained close to Paul and Emily Flory, and often saw them at Manchester College events. A warm letter of condolence and further correspondence was discovered at Manchester from Wylan to Emily on Paul's demise.

Paul Flory was also able to hire Thomas G Fox (Figure 9.4) as a research associate at Cornell. Many seminal papers resulted from this collaboration. Actual progress was made on the molecular weight dependence of the intrinsic viscosity. Staudinger was just wrong on this subject. Fox and Flory were able to use the new theory of the size of polymer chains as a function of solvent quality to show that the intrinsic viscosity scales as the square root of the molecular weight in a Flory solvent at the Θ point.[52] In a good solvent the intrinsic viscosity can scale up to the 0.8 power.

The melt viscosity of pure polymers also depends on molecular weight. For low molecular weight chains, the empirically observed dependence is linear. Above a certain molecular weight, the increase is as the 3.4 power. Fox and Flory published a classic paper on the viscosities of polystyrene[53] (Figure 9.5).

When the viscosity becomes so high that the liquid cannot achieve equilibrium in a convenient time, further cooling produces a nonequilibrium sample in the glassy state. Fox and Flory established an empirical relationship between the so-called glass transition temperature, T_g, and the molecular weight, M, for pure polystyrene[54]

$$T_g = 373 - 10^5/M$$

Many bulk polymers display this dependence.

FIGURE 9.4
Thomas G Fox (1921–1977). (From Carnegie Mellon University Archives. With permission.)

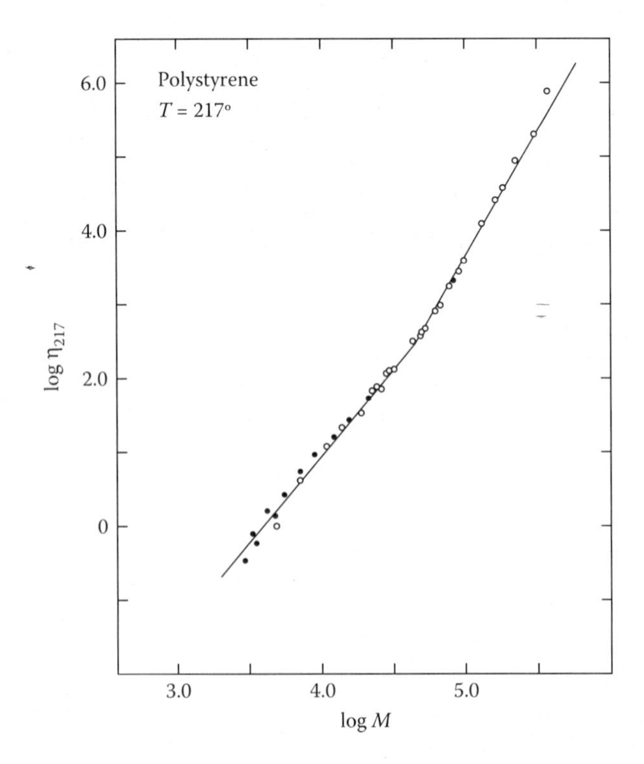

FIGURE 9.5
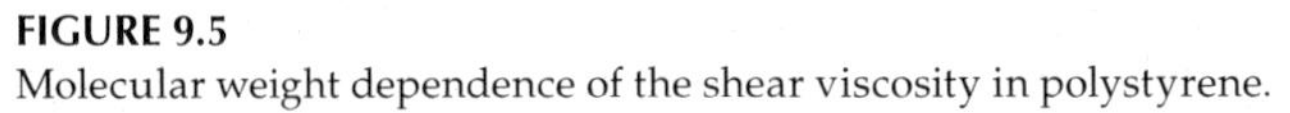
Molecular weight dependence of the shear viscosity in polystyrene.

A general theory of the intrinsic viscosity produced a very simple form[55]

$$[\eta] = \Phi \langle R^2 \rangle^{3/2} / M$$

where the prefactor is a universal constant, and the mean-squared end-to-end distance of the chain is used as the measure of the chain size. This formula accounts for all the laboratory observations, in contrast to the Staudinger Law, which accounted for none of them.

The immediate reward for Tom Fox was appointment as director of research at the Rohm and Haas Company in Philadelphia in 1950.

Another of the major figures in polymer science to collaborate with Flory at Cornell was Arthur M. Bueche (1920–1981) (Figure 9.6). He received his PhD in physical chemistry with Debye on the conformational and frictional properties of polymers in solution. He became an expert on light scattering from polymer solutions. A classic paper with Fox, Flory, and Bueche on osmotic and light scattering data appeared in the *Journal of the American Chemical Society* in 1951.[56] Bueche went on to become a star at the General Electric Company, a member of the National Academy of Sciences, and was nominated as the Presidential Science Advisor.

Cornell had many outstanding physical chemists in this period. Franklin Asbury Long (1910–1999) (Figure 9.7) contributed in many areas, including polymer science. He obtained his PhD from the University of California at Berkeley in 1935 with Axel Olson (1889–1954). He was a research associate at the University of Chicago with W.D. Harkins (1873–1951) and accompanied him to Cornell when Harkins was the Baker Lecturer in 1936–1937.

FIGURE 9.6
Arthur M. Bueche (1920–1981). (From National Academy of Sciences. With permission.)

Frank Long

FIGURE 9.7
Franklin Asbury Long (1910–1999). (From National Academy of Sciences. With permission.)

He remained at Cornell until the war years interrupted his tenure, but he returned in 1945 and was promoted to Full Professor in 1946. When Peter Debye stepped down as chemistry department chair in 1950, Franklin Long was appointed and served until 1960. He also served as vice president of Research and Advanced Studies from 1963 to 1969. He was elected to the National Academy of Sciences in 1962. In addition to his outstanding scientific and academic efforts, he was a tireless crusader for arms control. His firm stand led Richard Nixon to reject him as the director of the National Science Foundation. He joined Paul Flory as a leading member of the NAS group on human rights.

Another physical chemist and polymer scientist is Harold A. Scheraga (1921–) (Figure 9.8). He obtained his PhD from Duke University right after the war and was appointed to Cornell in 1947. He has devoted his career, which is still active, to the study of protein structure, dynamics and reactivity.

One of Flory's closest personal friends, Leo Mandelkern (1922–2006) (Figure 9.9), joined him at Cornell in 1949 as a research associate, after obtaining his PhD with Franklin Long. Mandelkern was blessed with the insights of Long, Debye, Kirkwood, Scheraga, and Flory. At Cornell, Mandelkern examined the use of frictional methods like diffusion, sedimentation, and intrinsic viscosity to characterize polymer chains in solution.[57] He went on to spend a decade at the National Bureau of Standards from 1952 to 1962. His specialty was the structure and properties of crystalline polymers. Leo and Paul stood shoulder to shoulder in many disputes on the nature of the

FIGURE 9.8
Harold A. Scheraga (1921–).

crystalline state in polymers. Leo then went to Florida State University for the rest of his career.

With the torrent of papers flooding from the Flory laboratory, the time had come to prepare his Baker Lectures for publication. Much new material had appeared since 1948, and considerable advances in theory had been made. The resulting book, *Principles of Polymer Chemistry*, was an instant success and is still in print! The unifying concept for the book is stated clearly in the preface: "The author has been guided in his choice of material by a primary concern with principles."[58] While the scope of the book is vast, especially for

FIGURE 9.9
Paul Flory, William Krigbaum, and Leo Mandelkern at Cornell. (From James Mark. With permission.)

1953, Flory was careful to note that "it would scarcely be possible in a single volume to do justice to all the excellent researches in the various branches of the subject."

With the appearance of his Baker Lectures and the other seminal papers from his Cornell years, Paul John Flory was elected to the National Academy of Sciences in 1953. He was very active in this group and led the fight for international human rights.

10

Flory at the Victoria University of Manchester

After a successful period at Cornell University, Paul Flory sought an opportunity to devote himself entirely to science on sabbatical leave. In 1953, his friend Geoffrey Gee (1910–1996) became professor of physical chemistry at the Victoria University of Manchester (Figure 10.1). This proved to be a great place to grow as a scientist. Paul Flory's sabbatical year was aided by a John Simon Guggenheim Memorial Fellowship. Gee himself had attended Manchester as an undergraduate and as a graduate student. He worked with D.C. Henry, the director of the Thomas Graham Colloid Research Laboratory, on electrocapillarity and received an MSc in 1933. Times were tough and Gee accepted a job with Imperial Chemical Industries. They immediately seconded him to Cambridge to work with Sir Eric Rideal (1890–1974). He was awarded the PhD in 1936. Another collaborator in the Cambridge period was Sir Harry Melville (1908–2000). Gee's background in physical and physical organic chemistry could not have been better.

Geoffrey Gee's first position was as a senior physical chemist at the British Rubber Producers Research Association (BRPRA) (1938–1953). He studied the properties of pure natural rubber and its solutions and gels with cross-linked rubber. He established a very high level of experimental precision; good enough to test available theories. The board of the BRPRA included such notables as Sir Walter Norman Haworth (1883–1950) and Sir Eric Rideal. They were pleased to have a fundamental research scientist at the Association. They also supported the appointment of L.R.G. Treloar (1906–1985). One of the crowning achievements of this period for Gee was election to the Fellowship of the Royal Society in 1951.[59]

The subdepartment of physical and inorganic chemistry at the University of Manchester was chaired by Michael Polanyi (1891–1976) from 1933, when he escaped from Europe, until 1948, when he shifted to the history and philosophy of science. The post was briefly held by M.G. Evans until his untimely death in 1952. The appointment of Geoffrey Gee turned out to be one of the best moves in the history of the University of Manchester. The department was very strong in the 1950s, with people like Sir John Rowlinson leading the study of liquids and solutions. Other polymer scientists included Colin Price, Colin Booth, C.E.H. Bawn, and Sir Geoffrey Allen. Geoffrey Gee went on to become the dean and pro-vice chancellor of the University of Manchester.

FIGURE 10.1
Geoffrey Gee (1910–1996). (From Victoria University of Manchester. With permission.)

He is even credited with seducing physicist Sir Sam Edwards into polymer science.

Detailed yearly reports of the chemistry department are extant.[60] The report for 1954 notes "We are very glad to welcome Professor Paul J. Flory who is on sabbatical leave from Cornell University. He has been elected an Honorary Research Fellow of the University and will make Manchester his headquarters during the coming session." Upon his departure the report for 1955 reports: "His visit was greatly enjoyed and very stimulating." Manchester was an important stop on the European chemistry circuit and many polymer scientists visited during that year. The report even notes that Henry Taube visited Manchester.

The scientific papers published by Paul Flory based on his work at Manchester reflect a deepening of his understanding of polymeric systems as examples of condensed matter physics.

The "new" theory of polymer solutions known as the Flory–Huggins theory assumed that the mixture was both positionally and orientationally random. Its success suggested that such an approximation was justified, at least in some cases. But, Paul Flory knew very well that more work needed to be done and was determined to do it while at Manchester. He extended the original lattice theory to include an energy-dependent term for chain flexibility. It was assumed that the straight conformation was the lowest energy, and that a particular chain would have an additional energy based on its tortuosity. He was able to show that for a pure liquid polymer, there existed a critical value of chain flexibility for an equilibrium liquid state. For systems of lower flexibility, one of the possible states for the system was the crystalline state. Flory and Mandelkern had been attacking this problem with vigor over the period from 1950 to 1955. While there are certainly intermolecular driving forces for crystallization, Flory was able to show that intramolecular effects alone predicted a phase transition to the ordered state.

He also considered the effect of dilution. When enough solvent is present, the system can remain disordered macroscopically. The phase diagram depends on chain length, chain flexibility and volume fraction of solvent. When the time came to submit the paper for publication, Geoffrey Gee sent it to the *Proceedings of the Royal Society*.[61]

When solutions of long rigid rods are more concentrated than a critical value, the solution phase separates into an isotropic dilute phase and an ordered concentrated phase. Lyotropic solutions have been known for a long time. Paul Flory decided to try and develop a theory of such solutions. He modified his usual lattice theory to allow for placement of rodlike entities. He sought the advice and help of theorists like John S. Rowlinson and applied mathematicians like S. Levine at Manchester. The result is still a classic. Flory predicted a form of the phase diagram that is still the basis for all further work (Figure 10.2). The paper immediately follows the previous one in the proceedings.[62] Flory came back to this problem many times, and was still working on it at the time of his death.

Geoffrey Gee continued his leadership role both at Manchester and in higher levels of government and was rewarded by being made Commander of the Order of the British Empire in 1958. He was lauded with a DSc, honoris causa, in 1983, from the Victoria University of Manchester. Paul Flory was also so honored in 1969, well before he received the Nobel Prize (Figure 10.3).

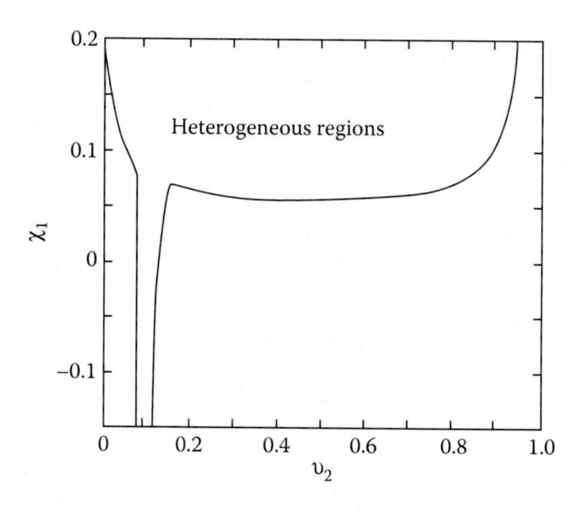

FIGURE 10.2
Phase diagram for a chain of length $x = 100$ as a function of solution interaction parameter and volume fraction of rods. (Flory, P.J., Statistical thermodynamics of semi-flexible chain molecules, *Proc. R. Soc.* London, A234, 60, 1956. Reproduced by permission of The Royal Society of Chemistry.)

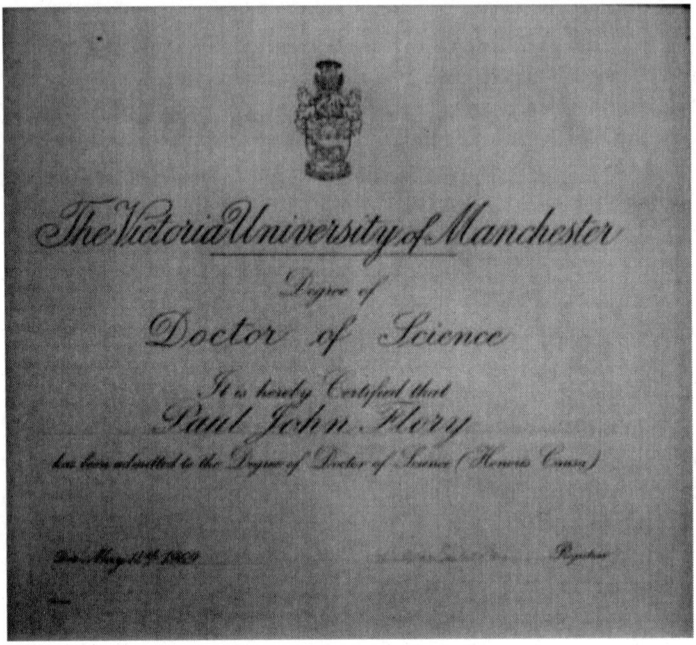

FIGURE 10.3
Honorary doctorate for Paul John Flory from the Victoria University of Manchester in 1969. (From Manchester University, [Indiana] Archives. With permission.)

11

*Flory at Mellon Institute**

The Mellon Institute for Industrial Research was founded in 1913 on the vision of Robert Kennedy Duncan (1868–1914) (Figure 11.1). Duncan was a chemical visionary who had written several famous accounts of the state of world chemistry in the early twentieth century. He initiated a program of industrial fellowships while he was a professor at Kansas. He believed that when industrial research was carried out in the same location as fundamental research, both groups benefited from the exchange of ideas.[63]

Two industrial pioneers in Pittsburgh, Andrew W. Mellon and Richard B. Mellon, desired to produce the same spirit of innovation in the chemical industry in America as had been demonstrated in Europe. They invited R.K. Duncan to come to the University of Pittsburgh as professor of industrial research in 1911 and to initiate the same program of industrial fellowships. A separate entity, the Mellon Institute for Industrial Research, was founded in 1913. Although Duncan did not live to see his ideas flourish, the Mellon Institute grew and prospered under the guidance of the Mellons. Much of the credit for the success of the enterprise can also be attributed to the leadership of Edward R. Weidlein (1887–1983) (Figure 11.2), who was a former student of R.K. Duncan at Kansas and joined the Mellon Institute in 1916. He served as its director from 1921 to 1956.

While practical searches for new polymers and new uses for polymers were well represented during the pre-War period, the time was not yet ripe for a specific fundamental effort in polymer science. The initiation of such an effort required a source of funds that could be devoted to pioneering research. The wartime need for the production of synthetic rubber provided an opportunity to eventually generate surplus money. E.R. Weidlein was very active in Washington circles and served on the board of the Reconstruction Finance Corporation. The Rubber Reserve Company was founded in 1942 to supply the country with synthetic rubber, and both academic and industrial firms were recruited for the effort. Weidlein made sure the Mellon Institute was also a part of this program. One of the great strengths of the Mellon Institute was in chemical instrumentation and a fellowship in synthetic rubber instrumentation was initiated to assist the development of online plant monitoring equipment.[42]

* This chapter is based on the paper "Polymer Science in the Mellon Institute" by Gary Patterson. It was presented to the regional ACS Meeting in Pittsburgh, PA in 2004.

FIGURE 11.1
Robert Kennedy Duncan, founder of the Mellon Institute. (From Carnegie Mellon University Archives. With permission.)

With the commercial success of the synthetic rubber program, the government now owned a very profitable business. Some of the proceeds of this enterprise were invested in pure research in polymers. In 1948 a pure research fellowship in synthetic rubber properties was founded in the Mellon Institute. The first member of this group was Thomas W. DeWitt. One of the

FIGURE 11.2
Edward R. Weidlein, director of the Mellon Institute (1921–1956). (From Carnegie Mellon University Archives. With permission.)

FIGURE 11.3
Hershel Markovitz, Society of Rheology Bingham Medalist. (From Carnegie Mellon University Archives. With permission.)

key hires was another Pittsburgher, Hershel Markovitz (1921–2006) (Figure 11.3), who had received his BS from the University of Pittsburgh in 1942 and his PhD from Columbia University in 1949. They formulated a program of research based on (1) careful synthesis of polymer samples, (2) chemical and structural analysis, and (3) precise measurements of the rheological properties of concentrated solutions and bulk materials.

The existence of well-characterized research samples of polymers encouraged fundamental studies using the best instrumentation. Infrared spectra of polyisoprenes were recorded and analyzed with W.S. Richardson in the chemical physics department. X-ray scattering was used to derive the radial distribution function for amorphous polymers with L.E. Alexander. By the end of 1951 the Synthetic Polymer Properties Fellowship had nine members and research was proceeding in many directions. Because the polymers produced could be rapidly characterized, systematic studies of polymerization were carried out as a function of temperature, pressure and light absorption.

In 1955 the United States Government decided to return the production of synthetic rubber entirely to private enterprise. The plants were sold and a unique era in polymer history came to a close. But, the acquired expertise of all the research groups supported by the Office of Rubber Reserve was considered a national asset and a temporary channel was created to continue funding this work. The polymer effort in the Mellon Institute was supported by a 2-year grant from the National Science Foundation, but the large group

FIGURE 11.4
Edward Casassa (right) and Yukiko Tagami (postdoctoral fellow). (From Carnegie Mellon University Archives. With permission.)

was reduced to two: Hershel Markovitz and Edward F. Casassa (1921–2009) (Figure 11.4). Casassa received his PhD from MIT with Walter Stockmayer and went on to become the editor of the *Journal of Polymer Science.*

The next period of growth in research in polymer science in the Mellon Institute was instituted by the appointment of Paul J. Flory (Figure 11.5) as the executive director of research in October of 1956. In order to compete in the new national environment it was decided to intensify efforts in fundamental research. Initial funding for the growth of this enterprise was provided by Mellon family funds, but it was anticipated that long-term support would come from various governmental and private agencies.

Paul Flory was one of the most visible polymer chemists and his monograph, *Principles of Polymer Chemistry,* provided a unifying perspective on a wide range of topics. He was a professor of chemistry at Cornell University and had been supported by the same office of Rubber Reserve. He was also one of the key scientists that met frequently to discuss progress during the war as part of the Rubber Reserve Corporation. He represented the Goodyear Tire and Rubber Company in these meetings. Weidlein knew him well. It was hoped that he could attract both a corps of outstanding scientists and the funds to support them. In the event, he did recruit an outstanding group of polymer scientists that persisted long after he departed, but a steady and ample source of funds for fundamental research did not materialize.

The most important appointment for the future of polymer science in the Mellon Institute was Thomas G Fox as assistant director of research. He met

MELLON INSTITUTE NEWS

Vol. XX Thursday, October 4, 1956 No. 1

WELCOME, DR. FLORY

Dr. Paul J. Flory
Our Executive Director of Research
(See pages 3-6)

FIGURE 11.5
Paul J. Flory. (From Carnegie Mellon University Archives. With permission.)

Flory at the Goodyear Company and they formed a formidable research team. He had worked in Flory's laboratory at Cornell from 1948 to 1950 and had established a successful industrial career at Rohm and Haas. He was a superb administrator and provided both scientific and organizational leadership. Many new fellows were appointed in emerging fields of chemistry. Flory created a small group that continued to work in areas of his direct interest. A.E. Ciferri (Figure 11.6) and C.A.J. Hoeve (Figure 11.7) studied rubber elasticity and established the extent to which changes in internal energy contributed to elastic properties. Hoeve also pursued the physical basis of muscle action. Flory's record of publication remained strong during his tenure at the Mellon Institute. There were many papers that were based on work done at Cornell. A series of papers on thermoelasticity was published based on the Mellon years. And a considerable number of papers on crystalline materials appeared. Flory was also a popular lecturer during this era.

FIGURE 11.6
Alberto Ciferri. (From Carnegie Mellon University Archives. With permission.)

His position at the Mellon Institute provided a solid base for his scientific efforts. However, administrative relations with the chairman of the Mellon Institute Board, Paul Mellon (1907–1999), became strained and Flory resigned as executive director of research in October 1960. He proceeded to Stanford University the following year.

FIGURE 11.7
C.A.J. Hoeve. (From Carnegie Mellon University Archives. With permission.)

FIGURE 11.8
Paul Flory, Paul Mellon, Tom Fox, and Edward Weidlein. (From Carnegie Mellon University Archives. With permission.)

The board of the Mellon Institute tried very hard to make Pittsburgh a good place for Paul Flory to live. The Mellon Institute leaders in 1957 are shown above (Figure 11.8).

12

Flory at Stanford: 1961–1985

The Leland Stanford Junior University opened its doors to both men and women in 1891. It was founded to be avowedly practical, producing cultured and useful citizens. It was built on the beautiful farm developed by Senator Leland Stanford near Palo Alto, California. The architect was Frederick Law Olmstead, the designer of Central Park in New York. The campus is still one of the most beautiful places in America (Figure 12.1). The first president was David Starr Jordan, the former president of Indiana University.

During the early years of the twentieth century, Stanford progressed as a regional university. Its most famous alumnus was Herbert Hoover. Good departments of science and engineering were established. After World War II, the pace quickened. The most notable individual from this era was Frederick Terman (1900–1982), the dean of Engineering (Figure 12.2). He is credited with the creation of the Stanford Industrial Park in 1951, where major Silicon Valley industrial firms like Hewlett-Packard worked in close collaboration with university researchers. It was a very synergistic relationship.

In 1955 Terman became the provost of Stanford University. During his tenure, Terman embarked on a campaign to build "steeples of excellence," clusters of outstanding science and engineering researchers who would attract the best students. He created an entrepreneurial spirit that still pervades Stanford University. The chemistry department wholeheartedly embraced this program.[64]

The renaissance of the chemistry department at Stanford started in 1960 with the appointment of William S. Johnson (1913–1995) as executive head (Figure 12.3). His mission was to recruit the best scholars in chemistry he could find. In December he learned accidentally that Paul Flory had resigned as director of the Mellon Institute, and within 15 minutes he obtained approval for an offer from Provost Terman. Not only did Stanford obtain Flory, but Henry Taube (1915–2005) (Figure 12.4) also agreed to come; the presence of Flory was noted by Taube as the deciding factor. The chemical trinity of Stanford was formed! Outstanding people like Eugene van Tamelen, Harden McConnell, Carl Djerassi, and John Baldeschweiler were added over the next 7 years. All seven new professors were members of the National Academy of Sciences. When I arrived in 1968, Stanford was a physical chemist's heaven. I was determined to work for Flory, but I took courses from all the other physical chemists, including the assistant professors, Hans Anderson and Robert Pecora. It was a good time to be at Stanford.

FIGURE 12.1
The Memorial Church, Stanford University.

FIGURE 12.2
Frederick Terman, provost of Stanford University (1955–1965).

FIGURE 12.3
William Sumner Johnson. (From Stanford University. With permission.)

When Bill Johnson retired from being head in 1969, Paul Flory agreed to serve for 2 years. He had already been acting head during Johnson's 1966–1967 sabbatical leave. This period was made especially difficult for Paul Flory by a deteriorating disk in his neck. I remember him sitting at a special vertical drafting desk so he would not need to bend his neck. He discharged

FIGURE 12.4
Henry Taube (Nobel Prize winner, 1983).

FIGURE 12.5
Seeley Mudd Chemistry Building. (From Stanford University Chemistry Department. With permission.)

his duties with enthusiasm and prodded the Stanford administration to plan a new and much larger chemistry building (Figure 12.5).

Rather than just rest on his previous accomplishments, Paul Flory struck out in new and renewed scientific projects. He laid down his marker on the question of the structure of semicrystalline polymers in a classic article in the *Journal of the American Chemical Society*[65] in 1962. Flory's arguments are clear, but they are based on concepts that largely assume some kind of equilibrium state. Flory and Mandelkern proposed that the lamellar structures observed in bulk crystallized linear polymers had diffuse interlamellar regions and that the morphology was quite complex. Other workers, thinking primarily of crystalline lamellae obtained from dilute solution, proposed a model where the predominant local chain structure was characterized by a high degree of adjacent reentry of the chains. In the event, very few crystalline polymer samples ever achieve anything like an equilibrium structure. Kinetic effects dominate the morphology, and the exact crystallization history is reflected in the observed structure. Like a stream, polymer crystals are never the same way twice. That does not mean there are no principles for this area, and Mandelkern carried out many experiments to establish the general factors that influenced crystallization. History has been much kinder to Flory's work in this area than to that of his adversaries.

James Mark joined Flory's group as a postdoctoral fellow and 23 years of fruitful collaboration began (Figure 12.6). The primary area of interest was rubber elasticity, but in order to understand this area, chain conformation must also be considered. Their first paper was on stress–temperature coefficients.[66] Specific studies of polymer configuration were carried out on poly(dimethylsiloxane),[67] poly(oxyethylene),[68] and poly(oxymethylene).[69] Further discussion is contained in Chapter 15.

Another of Paul Flory's former postdoctoral fellows and closest friends is Akihiro Abe (Figure 12.7). He obtained a PhD from Brooklyn Poly with Murray Goodman in 1963. The first paper with Flory on the thermodynamic properties of small molecule mixtures appeared in 1964.[70] An extensive

FIGURE 12.6
James E. Mark.

presentation of their long and warm scientific and personal history occurs below.

The time had also come to improve on the Flory–Huggins theory of liquid mixtures. Two collaborators of note in this effort are Robert Orwoll (Figure 12.8) and Agienus Vrij. An equation-of-state theory of n-alkanes and mixtures was developed and analyzed using literature data. It was a major advance over the early theory, but better data were needed over wider ranges in temperature and pressure. Orwoll carried out extensive measurements of

FIGURE 12.7
Akihiro Abe.

FIGURE 12.8
Robert Orwoll.

the equation of state for liquid n-alkanes and for n-alkane mixtures. Good values for the lower critical solution temperatures were obtained.[71] A personal reflection from Robert Orwoll is given below.

Agienus Vrij (Figure 12.9) joined the Flory group as a postdoctoral fellow in 1962 and stayed for a year, but his influence was felt throughout the work on liquids and liquid mixtures. He was appointed to The Utrecht University and went on to have a brilliant career in colloid and polymer science. A remembrance is found below.

While the Gaussian chain model of polymer chains is very useful, it contains no details of the specific macromolecule. A theory utilizing the rotational isomeric state model became popular in the 1960s and Flory and

FIGURE 12.9
Agienus Vrij.

FIGURE 12.10
Robert L. Jernigan.

collaborators pursued this area with vigor. Akihiro Abe joined this effort as well. The graduate student chosen to develop this theory was Robert Jernigan (Figure 12.10). Their first joint paper on second and fourth moments of the chain configuration appeared in 1965.[72] An extensive series of papers poured forth, culminating in 1969 with the publication of *Statistical Mechanics of Chain Molecules*.[73] A personal reflection from Robert Jernigan appears below.

Flory was also interested in biopolymers. David A. Brant (Figure 12.11) arrived as an NIH postdoctoral fellow in 1962. He studied the configurational statistics of polypeptides. Brant spent his academic career at the University of California at Irvine from 1965 to 2005, and served as the NSF program

FIGURE 12.11
David Brant.

FIGURE 12.12
Leonard Peller (with baby Brant, 1965).

director in biomaterials from 2006 to 2013. His area of special interest was the vast group of polysaccharides.

One of the scientists from the San Francisco area that was a frequent visitor to Stanford was Leonard Peller (1928–2000) (Figure 12.12). He received his PhD from Princeton and worked with Walter Kauzmann. He was a postdoctoral fellow at Wisconsin when Brant was a graduate student. And he was appointed to the University of San Francisco. He was a good theorist on biopolymers and could talk with Flory at the highest level. I always enjoyed Peller's visits.

In 1963, the liquid and solution group welcomed Bruce Eichinger (Figure 12.13). Like Robert Orwoll, he was both a good theorist and a careful experimentalist. After a year of postdoctoral work at Yale with one of America's

FIGURE 12.13
Bruce Eichinger.

FIGURE 12.14
Alan E. Tonelli.

best theorists, Marshall Fixman, Bruce went to the University of Washington. He continued to shine in both theory and experimental work throughout his career. A touching reminiscence is found below.

Another graduate student who joined the Flory group in 1964 was Allan E. Tonelli (Figure 12.14). He worked on a wide range of problems in configurational statistics, including vinyl polymers. He worked at AT&T Bell Laboratories in the polymer science department from 1969 to 1992 and joined the College of Textiles at North Carolina State University in 1992, where he is still active. A warm remembrance is included below.

The detailed study of the configurational statistics of biopolymers was a persistent theme throughout Flory's career. In 1967, one of his most successful theoretical students, Wilma King, entered Stanford (Figure 12.15). She has pursued the detailed understanding of polynucleotides ever since. She graduated from Stanford in 1971 as Wilma King Olson and has been a professor at Rutgers University ever since. A personal essay is found below.

The author entered Stanford in 1968 and promptly joined the Flory group (Figure 12.16). The laboratory was filled with old water baths and it was time to produce a new generation of instruments. He graduated in 1972, after a 2-year hiatus to satisfy the draft board, and joined Bell Laboratories in the Chemical Physics Department. A remembrance is found below.

Postdoctoral fellows continued to come from all over the world. From the University of Manchester, Chris Pickles joined the group and studied stereochemical equilibrium. A personal account occurs below.

FIGURE 12.15
Wilma King Olson.

As 1974 proceeded, there was a sense that big things were in the works. Paul Flory was awarded the Priestley Medal of the American Chemical Society at its spring meeting in Los Angeles. In light of all the other awards received by Paul Flory, it was no real surprise when it was announced that he had received the Nobel Prize in chemistry for 1974.

FIGURE 12.16
Nancy Lindner, PJF, and Gary Patterson.

13

Flory at Stanford: After the Nobel Prize

While the receiving of the Nobel Prize in chemistry in 1974 did not change Paul Flory in essential ways, it did change the level of his responsibilities outside of Stanford. From that point on he no longer accepted graduate students. A series of very successful postdoctoral fellows joined him, both at Stanford and in connection with the IBM Research Laboratory at San Jose. This chapter will focus on the scientific history of this period.

One of the other leading centers of polymer science in the United States was at the University of Massachusetts at Amherst. Richard Stein founded this effort in the early 1950s and it has thrived ever since. One of the best theoretical students ever to work for Dick Stein was Do Y. Yoon (Figure 13.1), and he chose to accept a postdoctoral fellowship with Flory. Not only was this joint effort fruitful, but when Yoon joined IBM Research, San Jose, a long-term collaboration was established. Elegant work in the theory of chain-vector distributions appeared almost immediately.[74] Treatments of advanced methods, such as x-ray and neutron scattering, for the analysis of polymer chain conformation were also presented.[75,76] In 1999, after 24 years of an IBM research career, Yoon moved to his alma mater as a full professor in the chemistry department of Seoul National University, Korea and upon retirement in 2012, returned to United States as a consulting professor of chemical engineering in Stanford University. A detailed personal narrative is included below.

Another major collaborator arrived in 1974 from the ETH Zurich, Ulrich W. Suter (Figure 13.2). He quickly joined the conformational analysis effort.[77] He made major progress in one of Flory's favorite areas: macrocyclization equilibrium.[78] Suter went on to have a brilliant career at MIT and returned as professor doctor to the ETH, finishing as vice president for research.

Paul Flory continued to attract postdoctoral fellows from all over the world. Burak Erman (Figure 13.3) came from Turkey. He focused on the theory of rubber elasticity while with Flory.[79] He was a professor in Bogazici University in Turkey for many years. He also collaborated with James Mark. An extensive narrative is found below.

Ever since his ground-breaking papers on polymeric systems with geometric constraints, Paul Flory worked on the theory of systems of rodlike particles. Lyotropic liquid crystals are an important part of polymer science. Georgio Ronca joined Flory in 1978 and produced more advanced theories in this area.[80] He had worked with Giuseppe Allegra at the Istituto di Chimica del Politecnico in Italy. He went to IBM after his time with Flory and collaborated with Do Yoon.

FIGURE 13.1
Do Y. Yoon.

The worldwide community of polymer scientists often shared their best people with Paul Flory. Bruno Zimm sent one of his very best, Ken A. Dill (Figure 13.4), to work with Flory in 1978. They collaborated on a project about membranes and micelles, another highly ordered system.[81] Dill went on to become a professor at the University of California at San Francisco. He became the president of the Biophysical Society. He was elected to the National Academy of Sciences in 2008. His personal remembrance is found below.

Another postdoctoral fellow from this period was Robert R. Matheson (Figure 13.5). He came from Harold Scheraga's lab at Cornell. He worked

FIGURE 13.2
Ulrich W. Suter.

FIGURE 13.3
Burak Erman.

with Flory on the physics of liquid crystalline macromolecules. He took this expertise to DuPont, where he has had a brilliant career.

Although Paul Flory got older, he did not slow down. His scientific productivity continued vigorously throughout his AARP (American Association of Retired Persons) years. His major activity during the period from 1974 to 1985 will be described in the chapter on his humanitarian work.

FIGURE 13.4
Ken A. Dill.

FIGURE 13.5
Robert R. Matheson.

A birthday celebration was held for Paul Flory's 75th anniversary. Hundreds of friends gathered at Stanford for the event. There were the usual ceremonial talks, but the focus was on the continuing friendship with so many people (Figure 13.6). The other attendees included most of the people in the Polymer Science Hall of Fame. The symposium program included many of these friends (Figure 13.7).

FIGURE 13.6
Group photo at Paul Flory's 75th birthday party. (From Stanford Chemistry Department. With permission.)

FOUNDATIONS AND CHALLENGES OF POLYMER SCIENCE

CELEBRATING PAUL FLORY'S 75th BIRTHDAY

Monday, June 17	Registration–Reception 7–9 pm—Buck House Shuttle van departing from Seely Mudd Chemistry Bldg at 6:30 p.m.	
Tuesday, June 18 8:00–8:30	Registration–Coffee at Braun Auditorium	
8:30–9:30	Session A—Thermodynamics and Hydrodynamics	
	Moderator:	A.R. Shultz
	Speaker:	W. Stockmayer—30 min.
	Discussants:	M. Fixman, R.A. Orwoll, R. Pecora, B.H. Zimm
9:30–10:45	Session B—Chain Conformation General Principles	
	Moderator:	P. Pino
	Speaker:	W.L. Mattice—30 min.
	Discussants:	A. Abe, J.A. Semlyen, U.W. Suter, A. Tonelli
10:45–11:00	Coffee Break	
11:00–12:15	Biopolymers:	
	Moderator:	F.A. Bovey
	Speaker:	H.A. Scheraga—30 min.
	Discussants:	D.A. Brant, K.A. Dill, R. Jernigan, W.K. Olson,
12:15–1:30	Box Lunch on Terrace	
1:30–2:00	Session B—Chain Conformation Chain Dynamics	
	Moderator:	J.D. Ferry
	Speaker:	E. Helfand, S. Edwards
2:00–3:00	Session C—Amorphous Chain Structure	
	Moderator:	C. W. Frank
	Speaker:	E.W. Fischer—30 min.
	Discussants:	H. Benoit, G.D. Patterson, D.R. Uhlmann
3:00–3:15	Coffee Break	
3:15–4:30	Session D—Rubber-Like Elasticity	
	Moderator:	A. Cifferi
	Speaker:	J.E. Mark—30 min.
	Discussants:	G. Allegra, B.E. Eichinger, B. Erman, T.L. Smith
6:30	Wine Bar and Banquet at Faculty Club Wine Bar—6:30 and Dinner—7:30	

FIGURE 13.7
Program for Flory 75th birthday symposium.

(*Continued*)

Wednesday, June 19

8:30–9:45	Session E—Crystalline Polymers	
	Moderator:	J.G. Fatou
	Speaker:	L. Mandelkern—30 min.
	Discussants:	E.W. Fischer, G.D. Wignall, D.Y. Yoon
9:45–10:00	Coffee Break	
10:00–11:15	Session F—Liquid Crystals	
	Moderator:	J.R. Schaefgen
	Speaker:	W.R. Krigbaum—30 min.
	Discussants:	J. Economy, R.R. Matheson, W.G. Miller, A.H. Windle
11:15 -	Concluding Remarks	
	Moderator:	J. E. Mark, N.M. Bikales, D.R. Ulrich
	P.J. Flory—Introduced by H.F. Mark	

FIGURE 13.7 ***(Continued)***
Program for Flory 75th birthday symposium.

While Paul Flory enjoyed his travels and his public moments, he often felt his best when he was at work in the midst of nature. Even when he was at his retreat at Big Sur, he was often hard at work (Figure 13.8).

On one of these weekends, he made his last discovery (Figure 13.9). A memorial service was held on September 23, 1985 (Figure 13.10).

Throughout this period in Paul Flory's life, James Economy (Figure 13.11) played an important role. He was manager of the Polymer Science and Technology Department in the IBM Research Laboratory in San Jose. He provided an additional scientific home for Flory and supported postdoctoral fellows.

Although Paul John Flory was gone, he was not forgotten. He received the ACS Polymer Chemistry Division Award in Polymer Education posthumously in 1986. The 3 days of symposia were a fitting wake for his friends (Figure 13.12).

Paul John Flory was inducted into the International Rubber Science Hall of Fame on November 7, 1986. Maurice Morton (1913–1994) (Figure 13.13), his longtime friend from the University of Akron, and James E. Mark wrote the eulogy article.[82] The official citation reads: "Paul John Flory (1910–1985), American chemist, truly the 'founder of polymer science,' who, in his wide-ranging investigations, both theoretical and experimental, laid the foundations for the science of macromolecules, including major contributions to the theory of rubber elasticity." His official portrait hangs in the University of Akron (Figure 13.14), along with so many of his friends!

FIGURE 13.8
Paul Flory at work in Big Sur. (From Susan Flory Springer. With permission.)

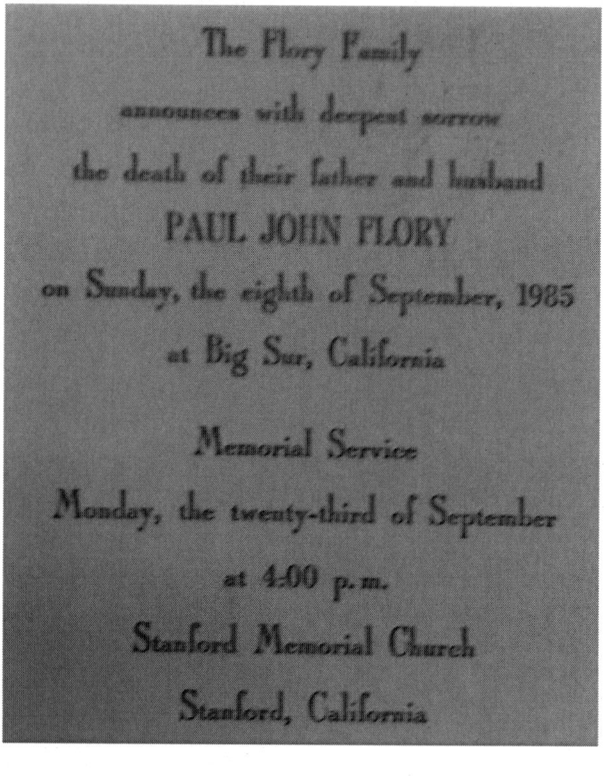

The Flory Family
announces with deepest sorrow
the death of their father and husband
PAUL JOHN FLORY
on Sunday, the eighth of September, 1985
at Big Sur, California

Memorial Service
Monday, the twenty-third of September
at 4:00 p.m.
Stanford Memorial Church
Stanford, California

FIGURE 13.9
Memorial announcement. (From Manchester University Archives. With permission.)

Stanford Memorial Church

September 23, 1985 *4:00 p.m.*

Organ Prelude *Prelude and Fugue in E^b major* J.S. Bach (1685-1750)

Introductory Prayers, Remarks and Reading

Tributes

Dr. Morris Pripstein
Senior Physicist, Lawrence Livermore Laboratory, Berkeley

Dr. Ken A. Dill
Associate Professor of Pharmacological Chemistry and Pharmacy, University of California San Francisco

Dr. James Economy
Manager of Polymer Science and Technology, IBM Research Center, San Jose

Closing Prayers

Organ Postlude *Sinfonia from "Wir danken dir, Gott"* J.S. Bach

Minister: Ernle W.D. Young, Associate Dean of the Chapel and Chaplain to the Medical Center

Organist: Robert Bates, Assistant University Organist

FIGURE 13.10
Program from memorial service for Paul Flory. (From Ken Dill. With permission.)

FIGURE 13.11
James Economy.

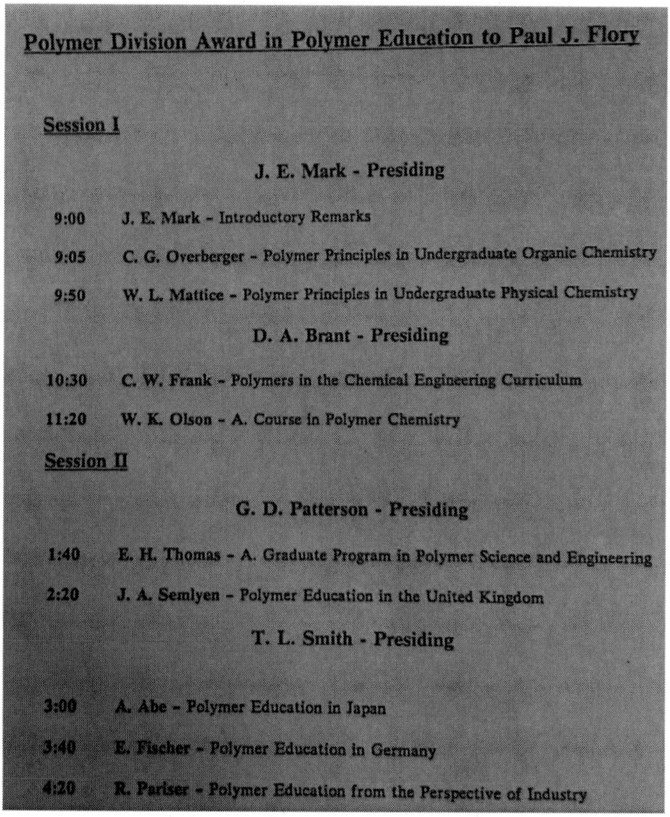

Polymer Division Award in Polymer Education to Paul J. Flory

Session I

J. E. Mark - Presiding

9:00 J. E. Mark - Introductory Remarks

9:05 C. G. Overberger - Polymer Principles in Undergraduate Organic Chemistry

9:50 W. L. Mattice - Polymer Principles in Undergraduate Physical Chemistry

D. A. Brant - Presiding

10:30 C. W. Frank - Polymers in the Chemical Engineering Curriculum

11:20 W. K. Olson - A. Course in Polymer Chemistry

Session II

G. D. Patterson - Presiding

1:40 E. H. Thomas - A. Graduate Program in Polymer Science and Engineering

2:20 J. A. Semlyen - Polymer Education in the United Kingdom

T. L. Smith - Presiding

3:00 A. Abe - Polymer Education in Japan

3:40 E. Fischer - Polymer Education in Germany

4:20 R. Pariser - Polymer Education from the Perspective of Industry

FIGURE 13.12
Symposium program for the education award of Polymer Chemistry Division of the American Chemical Society (POLY).

FIGURE 13.13
Maurice Morton. (From International Rubber Science Hall of Fame. With permission.)

FIGURE 13.14
Paul John Flory (1910–1985). (From Stanford University Archives. With permission.)

14

Emily Catherine (Tabor) Flory: Ultimate Partner

Emily Catherine Tabor was born on January 16, 1912 in Boyertown, Pennsylvania. Her father, Leon Raymond Tabor (1891–1965), had been born in Pennsylvania, and by the time of her birth had settled in Berks County. Her mother, Amy Alberta (Kurtz) Tabor (1893–1986), outlived her husband and her son-in-law.

Emily graduated from Boyertown High School in 1929. Although she was an accomplished classical violinist, she chose to go to Drexel Institute and major in home economics. She graduated in 1933 at the height of the depression, but she obtained a job with Delaware Power and Light in Wilmington, Delaware, the home of DuPont. Emily specialized in interior lighting design and became known in Wilmington. She met Paul Flory at a party held by DuPont to welcome new employees.

Paul and Emily married in 1936 in the Washington Memorial Chapel in Valley Forge, Pennsylvania. They welcomed their first child, Susan, on August 20, 1938. Soon afterwards it was time to leave Wilmington and settle briefly in Cincinnati (1938–1940). This was back in "Flory" country, and they visited family, friends, and favorite places, like Manchester College and Carl Holl.

Emily needed to pack lightly, since Paul moved on to New Jersey to work for Esso Laboratories. While the time in New Jersey was also brief, they made friends with John Rehner, Jr. Their second child, Melinda, was born on October 19, 1940 in New Jersey.

A new job at Goodyear Tire and Rubber Company in Akron allowed a return to Ohio in 1943. The years at Goodyear were busy and productive. Paul John, Jr. was born on April 21, 1946.

Appointment as professor of chemistry in Cornell University meant settling in Ithaca, New York. This was a very happy time with family outings to Lake Cayuga for sailing and picnics. Emily was very active in the local Unitarian Church. While there may be many differences between Unitarians and the Church of the Brethren, they were both committed to peace and justice.

The sabbatical year in Manchester, England was enjoyed by the whole family. They lived in the country in a village called Chinley, Derbyshire, on the edge of the moors.

The position as research director of the Mellon Institute required substantial participation of the Director's wife. Formal teas were served in the

FIGURE 14.1
Flory home on Buckingham Road, Fox Chapel, Pennsylvania.

elegant social room of the Mellon Institute and Emily was the gracious hostess. A beautiful home in the hills of Fox Chapel was also purchased (Figure 14.1).

The move west to Stanford also meant another beautiful home in Portola Valley. One of the traditions of the Flory laboratory was the yearly party at the Flory home. Everyone was also invited to swim in the pool; Paul and Emily used it regularly and were in top shape. Another tradition was Emily's lecture to the spouses of the graduate students. Paul and Emily enjoyed camping and hiking in the Sierras.

They built a cabin in Big Sur that served as a retreat (Figure 14.2). Emily continued to visit the site until it became a burden to keep it up.

When Paul Flory won the Nobel Prize in 1974, Emily was by his side, as always (Figure 14.3).

FIGURE 14.2
Flory cabin at Big Sur, California.

Emily was also with him when he visited Manchester College in 1975. She became very good friends with the wife of the President, Pat Helman (Figure 14.4).

After the Nobel Prize, Emily and Paul were often traveling. Paul Flory was given many honorary doctorates and invited to give many plenary lectures.

FIGURE 14.3
Emily and Paul. (From Manchester University Archives. With permission.)

FIGURE 14.4
Emily and Pat. (From Manchester University Archives. With permission.)

He was also especially active in promoting human rights. Emily was a full partner in this effort.

After Paul Flory's death, Emily soldiered on. One of her first major tasks was to donate Paul's medals and awards to Manchester College (Figure 14.5).

Another major gift was 72 boxes of letters and other papers given to the Center for the History of Chemistry (now the Chemical Heritage Foundation). This biography has benefitted greatly from these archives.

FIGURE 14.5
Emily and President Helman. (From Manchester University Archives. With permission.)

Although all the other bound theses of Flory's students at Stanford are located at the University of Akron (along with a collection of personal scientific books from Paul's library), Emily sent my red-bound copy to me directly.

Emily Flory was a strong woman with firm principles. She devoted herself to promoting Paul's career. She was unfailingly supportive and gave excellent advice. She will be remembered by all of Paul Flory's many friends. She lived to be 94, just like her mother!

15

Friends of Flory

CONTENTS

One of the benefits of a biography written close in time to the death of the subject is that many of the people who knew him well are still living. Many of Paul Flory's students and postdoctoral fellows from the Stanford era have contributed personal narratives. They have been lightly edited, but the intention is to "hear" them in their own voice.

James E. Mark

Jim Mark was limited with regard to his choice of which college to attend for an undergraduate degree, since his family had little funding for this purpose. He therefore picked Wilkes College, a local school that started out as part of Bucknell Junior College. The choice there was between chemistry and music (as a clarinetist), and the former was chosen since it was a 4-year program, while the latter was for only 2 years at that time.

Jim took a year off during this time for financial reasons, and ended up working at Rohm and Haas in Philadelphia. This was a fortunate choice since Tom Fox had recently joined the company from Flory's group to direct research on acrylate and methacrylate polymers. Tom Orofino, who had also worked with Flory, had also been there as part of this polymer group. A major goal was the preparation of isotactic poly(methyl methacrylate) (PMMA) and its success was an important fundamental achievement, but one of little practical or commercial importance. Unfortunately, this may have had a bad effect on attitudes to basic research there, in general. In any case, these interactions with Tom Fox inspired Jim to stay in the polymer area for the rest of his career.

Jim got his PhD from the University of Pennsylvania in 1964, under Bob Hughes. Bob was trained as a crystallographer, at Cornell, under Lynn Hoard. When Bob and his wife moved to Philadelphia, she got a job at Rohm and Haas, and this got Bob to add polymer chemistry to his list of interests. This is the area in which Jim worked, specifically the effects of stereoregularity on the solution properties of poly(isopropyl acrylate). During this period, Paul Flory gave a lecture at Penn, impressing many in the audience, including Jim. Jim wrote to Paul Flory about a postdoctoral position, and was offered one, to be held at Stanford University, since Paul was moving there from the Mellon Institute in Pittsburgh.

Brian Jackson was working with Flory on polymer crystallinity at this point and he was instrumental in helping with the move to Palo Alto. After doing some additional research at Stanford, Brian returned to England, to take a job at the large industrial firm Industrial Chemical Industries (ICI). He stayed there until he was let go in a reorganizing/downsizing. The group, as it existed in the early 1960s, is shown below (Figure 15.1).

Paul Flory and Tony Semlyen are deceased, and Bill Gorth and Steve Fisk apparently left the polymer area. In fact, when Jim was sending out invitations for a Flory party, he found that only about one-third of all the Flory colleagues stayed in the area of basic polymer research!

Jim first started working with Paul Flory in the area of the spatial configurations of polymer chains, using rotational isomeric theory. Specific examples would include poly(dimethylsiloxane) (with Vittorio Crescenzi), various polyoxides (with Akihiro Abe and Sidney Bluestone), and several stereochemical forms of vinyl and related polymers (with Akihiro Abe). The properties of particular interest were unperturbed dimensions and dipole moments. The case of the dimensions of isotactic polypropylene proved to be a very contentious issue. The samples in hand were said to be nearly 100% isotactic, on the basis of NMR (nuclear magnetic resonance) results, by Frank Bovey and others. Rotational isomeric state calculations indicated that such a polymer would have very large dimensions in solution, because of the high isotacticity forcing long-chain sequences into helical arrangements. Experiments showed that the dimensions were not unusually large. The NMR group attributed this to shortcomings in the rotational isomeric

FIGURE 15.1
(Reading from left to right) Paul Flory, Jim Mark, Bob Jernigan, Bill Gorth, Dave Brant, Akihiro Abe, Steve Fisk, Bruce Eichinger, and Tony Semlyen.

state calculations, and the computational group attributed it to the NMR results overestimating the percentage of isotacticity. Modifications in both approaches brought them into closer agreement, and this is no longer a divisive issue. There was extensive litigation, however, over the patents on the preparation of the isotactic form of polypropylene, and this lasted a number of years! The other major area investigated was the stress–strain isotherms of elastomers, particularly poly(dimethylsiloxane).

Jim returned several additional times to work further with the Flory group, at Stanford in 1973–1974 and at Stanford/IBM-San Jose in 1984–1985. The projects in these cases focused on various aspects of rubberlike elasticity.

Paul almost always looked serene and full of self-confidence, but he did have a temper. Apparently he fired his first student at Cornell, and his first student at Stanford. He also had little patience with people who disagreed with any of his ideas on scientific topics. There was once a conference at Stanford on "Order in the Amorphous State in Polymers," and it was supported by a major funding agency. Paul was convinced that there was no such order, and apparently no one was invited to the conference who thought otherwise. Also, he was not celebrated for his tact and diplomacy in arguments that took place at a number of conferences.

Jim and Paul remained close friends until Paul's unfortunate passing in 1985, which represents a period of more than 20 years. Their friendship even included having Jim and his wife house-sit the Flory home, when both Paul and his wife Emily were both on extended trips outside the country.

Akihiro Abe

What Science Is All About: Recollections from the Days with Paul J. Flory (PJF)

1964–1965

How I Met Him

Paul Flory was always very quick in responding to others. In 1963, I was a PhD student just finishing up my thesis under Professor Murray Goodman (MG) at the Polytechnic Institute of Brooklyn, and looking for a postdoctoral position for some additional years. My first contact was Dr. A.J. Kovacs (AJK) of CNRS (Strasbourg). On one hand, MG got in touch with PJF as the second choice. After a couple of steps, I received an offer on June 14 from PJF. The delayed reply from AJK (signed by Professor H. Benoit) arrived in July, a little too late. This was the start of my long association with PJF. I left Brooklyn-Poly in September to visit Europe and eventually arrived in Tokyo at the end of October. Somehow I succeeded in convincing my company (Showa Denko Co.) to allow me to extend my stay in the United States. Masako and I, a newly married couple, arrived in Palo Alto early in February 1964 (Figure 15.2).

Later AJK made a joke on me "If my letter reached you earlier, I might be the one to get a Nobel Prize".

The Flory group was close knit in 1965 and many of the people from this time went on to have great careers in polymer science (Figure 15.3).

FIGURE 15.2
A home party at the Portola Valley residence in 1965.

FIGURE 15.3
The Flory group at Stanford in 1965.

Gordon Research Conference in 1965

The Gordon Research Conference held (on July 5–9) in 1965 was one of the epoch-making meetings in polymer history. About 140 attendants got together at Colby Junior College, New Hampshire. As I remember, PJF and Stocky (Walter Stockmayer, ed.) had a hot debate on their hydrodynamic expressions in front of a big audience: PJF was furiously tough on the opponent.

1966

IUPAC Macro in Tokyo and Kyoto

The IUPAC Macromolecular Symposium was held in Tokyo and Kyoto for the first time. Jim Mark and PJF presented a paper on the rubber elasticity in a general session. At a banquet held this time or another occasion in these years, Professor W. Stockmayer (Stocky) asked me if I have ever heard of the expression "Flory-watcher." He explained it to me mentioning Flory's amazingly high intuitive power. Yes, PJF exceeded others in both intuition and power of concentration.

After the symposium, PJF and Emily spent one evening with us at our house in a suburban area of Tokyo. They enjoyed Japanese dishes while sitting on the Japanese-style "tatami" mat (Figure 15.4).

1971

Lecture Tour in Japan

In May, PJF was invited to deliver a plenary lecture at the meeting celebrating the 20th anniversary of the Society of Polymer Science, Japan (SPSJ). On this

FIGURE 15.4
PJF and Emily as the guests at our house, with my father on the left.

occasion he made extended visits to major universities in Tokyo, Kyoto, and Osaka. His lectures were attended by a great audience.

An Embarrassing Happening

During the coffee break after the lecture at Osaka University, two students came in and respectfully asked him to sign his recent book. The copies they presented were apparently pirated editions. PJF *embarrassingly* glanced at me, and signed them like he had not noticed. Later they were scolded heavily by their supervisor. Of course no more such editions exist nowadays (Figures 15.5 and 15.6).

FIGURE 15.5
Lecture on polymer thermodynamics at Kyoto University in 1971.

FIGURE 15.6
PJF holding my son at the departure from Narita Airport in 1971.

1975

The Second Extended Stay at Stanford in 1975

In 1974, PJF received the Nobel Prize. In the succeeding year, he invited me as a visiting scholar to stay at Stanford. My company allowed me to take a year off on leave from my duty. Mrs Emily Flory kindly met us at the airport and took us to a cute house previously arranged in Menlo Park. This was the start of my second extended visit with PJF at Stanford. It was an extremely gorgeous time in my life. I was allowed to use my time just for myself without any intervening obligation. After the prize, PJF became quite busy and he was often out of town. Nevertheless, he quite often spared time for me to talk about various topics of common interest. I enjoyed being a "Flory-watcher" (Figure 15.7).

On June 17, I participated in the symposium honoring his Nobel Prize held at Carnegie-Mellon University: for the first time I saw him making a joke in his speech at the dinner.

In the summer, we carried out a transcontinental family tour in our tiny Toyota Corolla. In Boston we visited PJF at MIT: he was about to decide not to move to the east after his retirement from Stanford University.

FIGURE 15.7
PJF and Emily with my family at Big Sur in January 1976.

In the fall, we visited their country house at Big Sur (Figure 14.2). PJF invited me to go for a walk around the hill, full of ups and downs. Whenever we came to a branched uphill road, he said "would you mind taking the steeper path?" PJF took out a telescope and explained to our children how to watch whales passing by along the coast.

A Lifelong Friendship with Maurice L. Huggins

In the lecture entitled "Concepts in Polymer Science, a Half Century in Retrospect" delivered at the IUPAC meeting (1985), PJF mentioned the rivalry relation with Maurice L. Huggins (MLH): Upon learning of my efforts in the same area, Huggins magnanimously encouraged me to publish my work concurrently with his. This was the start of a lifelong friendship, unmarred by rivalry over priority.

MLH was about to retire from the Stanford Research Institute (SRI) according to their internal rule. One hot day in 1975, PJF visited the top management of SRI to ask for an extension of their appointment to MLH by explaining his important role in polymer science. A couple of hours later PJF came back with a stern look, shaking his head gently from side to side. They are nuts!

1982

Three Months at IBM and Stanford

PJF and Do Yoon invited me to stay at IBM San Jose Research Lab and Stanford for half a year starting from April. We rented a cottage on the Stanford campus. Our children enjoyed the summer program at Menlo School (Figure 15.8).

FIGURE 15.8
A home party at the Portola Valley residence in 1982.

This year the IUPAC Macromolecular Chemistry Congress was held at University of Massachusetts (UMass). As I remember, the campus was filled with a large number of mosquitoes.

Hostage

PJF was very sensitive to the human-rights movement. It was rather widely known that PJF wrote a letter to the Soviet leader to free Andrei Sakharov and his wife to get medical treatment in a western hospital. In the letter, he stated that if they require a hostage in exchange, PJF would willingly volunteer. He was really serious.

Science is But a Game!

Sometime in 1982, although I do not remember exactly when, Flory clearly mumbled "*science is but a game*!" while chatting in the laboratory at Stanford. We hit it off right away.

The Kigetsu-Shoin Foundation

After my father, I have been running a scholarship foundation in my country. The organization has been providing financial support to university students, coming from all fields from natural to human sciences, to encourage them to build up their human character. Each year the third-grade students are assigned to organize a seminar of high quality. Old timers join to study together. The progress of our program was often the lunch-time topic at the IBM cafeteria. He used to tell me about activities of similar organizations which attracted his attention worldwide. I will always be grateful for his friendship and encouragement.

1985

The Symposium "Foundations and Challenges of Polymer Science, Celebrating Paul Flory's 75th Birthday" (at Stanford)

As we know, this became the final page in his scientific career. I arrived in San Jose on May 30, after spending some time in New York and San Diego. I was happy to see PJF in good health. In fact, he was strong enough to mow the grass by himself in the front and back yards of the house in Big Sur.

The reunion meeting for the anniversary was held on June 17–19, attended by about 150 people. Members of the Flory school renewed their old friendship after many years. At the Banquet, "Selected Works of Paul J. Flory" (edited by L. Mandelkern, J. E. Mark, U. W. Suter, and D. Y. Yoon) was presented. On behalf of the Chemical Society of Japan, I presented the certificate of honorary membership (Figures 15.9 and 15.10).

Herman Mark chaired the last session of the symposium. PJF stood up and gratefully talked about how much he enjoyed collaborating with various people through science, and mentioned prominent examples developed by his former associates after Stanford. As the last speaker, he firmly concluded the session by declaring "Let's get back to work!" (Figure 15.11).

JAL Flight 123 Crashed to the Mountain

In the evening of August 12, he dropped in my office at IBM, and told me that a jumbo-jet crash had taken place in Japan: 520 people died. Thanks to his

FIGURE 15.9
A big reunion on the Stanford campus in 1985.

FIGURE 15.10
The feelings of nostalgia (with Tony Semlyen).

quick communication, I was immediately able to make sure nobody I knew was involved in the accident.

The Plenary Lecture at IUPAC Macro

As mentioned earlier, PJF was asked to deliver a lecture on the history of polymer science at the IUPAC Macromolecular Symposium in The Hague,

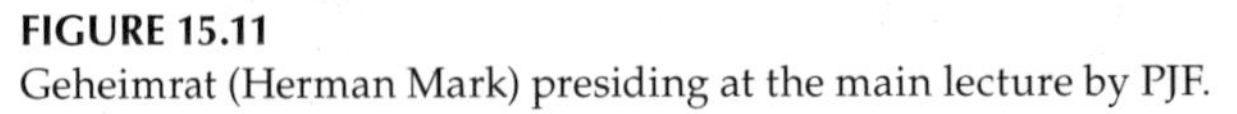

FIGURE 15.11
Geheimrat (Herman Mark) presiding at the main lecture by PJF.

FIGURE 15.12
Notes of the last discussion with PJF during lunch at the faculty club.

August 18–23. He said to us "the organizers were forcefully asking me to accept, but I think this is not the right time to look back." In fact, he was very reluctant as if he seemed to have foreboded his own destiny. Somehow PJF managed to handle his busy schedule in this spring and summer. His health looked OK except for some occasional coughing and chest pain.

Farewell to PJF

On August 30, PJF and I had our last lunch at the Stanford Faculty Club. I remember the topic at the lunch (Figure 15.12). We walked back to his office. When I said goodbye, he made two closes by using the first and second fingers of both hands. I responded to him "Please don't, you should remember that we are making plans for your 80th birthday." The next day I flew back to Tokyo. One week later (September 8), I received a phone call at my house from his secretary "I have sad news." She continued "Flory himself." I lost words. Emily's message followed "He was dead. Don't have to try to attend the funeral ceremony from far distance."

PJF was a man of great intuition and concentration. Above all, he was a great human. Later, on another occasion, Emily emotionally put it "This may be a right time for Paul. Perhaps he could not stand for his weakening steps."

Epilogue

1986

ACS Polymer Education Award to PJF

Polymers are one of the most complex systems of matter. PJF always emphasized the importance of fundamental studies in the field. At the interview by

Ridgway, he left the famous phrase "The field should attract to it students and faculty with more abstract and basic science orientations. Neglect in this respect, I think, will not only be to the detriment of polymer chemistry, but to the detriment of chemistry as a whole. The field of chemistry cannot afford to abandon a sector of such widespread relevance." (*J. Chem. Educ.*, 54, 341–344, 1977).

After the memorial symposium in Honor of PJF at the ACS meeting in Anaheim, the Polymer Education Award was presented posthumously to PJF from the Polymer Chemistry Division.

1994

Flory behind the Journal, Macromolecules

In 1994, F. H. Winslow, the chief editor of Macromolecules distributed his recollection "Macro 25" at the Editorial Advisory Board meeting for leaving members (the ACS spring meeting in San Diego). There again I have learned of the initiative of PJF in the journal's history. According to Winslow, "The journal was conceived in January, 1963 when Paul Flory wrote to Frank Mayo, then Head, Committee on Macro. Chem. at the NRC, asking the committee to make an appraisal of existing polymer publications. Flory was unhappy with existing journals in general. Three years later in 1966 the NRC Committee concluded that the ACS needed to have its own polymer journal."

2003

Memorial Symposium "Foundations of Polymer Science, Paul J. Flory's Seminal Contributions, Present Status, and Projections" was organized at the ACS Fall Meeting in New York (Figures 15.13 and 15.14)

Old friends and the former associates of PJF got together on this occasion. At the conference and dinner table, people recaptured the past variously. Apparently PJF had mellowed as time went by. After all, he showed us what (classical) science was all about. Throughout their life, Paul and Emily loved nature (particularly the red woods) and Paul prioritized humanity of all other things.

2006

Farewell to Emily

In 2007, I received a letter from Mrs. Susan Springer, Flory's elder daughter, informing us that Emily passed away on April 3, 2006 at the age of 94. Susan wrote "She continued to live by herself until almost the very end. As you know she was an amazingly strong and independent woman." She shared her entire life with Paul. I am sure that she must have lived for Paul throughout her life even after his death.

FIGURE 15.13
The reunion at the Flory Symposium in New York in 2003.

Concluding Remarks

In my life, I have had many more contacts with PJF on various occasions other than those mentioned above. Whenever I saw him, I learned something new. Because of the limitations of space, I just picked important incidents for the purpose at hand. Those in 1974, 1977, 1984 also remain fresh in my memory.

Now almost 30 years have gone since PJF left us. Today we find a quite different atmosphere surrounding scientists in academic institutions. As pointed out everywhere, while modern scientific instruments have become more expensive, the number of scientists has increased tremendously worldwide. As a result, competition for financial support is now extremely tough. Under these circumstances, projects pursuing immediate profits tend to be preferentially funded. In these technology-oriented projects, younger scientists are often required to collaborate with each other in a strictly organized manner. Faculty positions are granted by

FIGURE 15.14
After dinner gathering in 2003.

the productivity (the number of publications and the impact factor) of the applicant. On one hand, major concerns raised for science such as those of the environment are of complex nature, and elucidation of the mechanism often requires an extended cooperation of many workers including those from different disciplines.

Will Flory-style science then come back again in the near future? The answer is quite negative. As long as science continues, however, there should be a need for innovative ideas evolved in an unexpected way. These do not come up as a result of group discussions. As suggested in the history of science, personal intuitions are an indispensable source of progress. Younger scientists must have a strong motivation to pursue their own ideas. They may need time to relax their minds from their targeted teamwork. Let us hope that they will overcome the difficulties they encounter.

Robert Orwoll

First Meeting with Paul Flory

In 1962 I was admitted to the graduate program in chemistry at Stanford University. Paul Flory had joined the Stanford chemistry faculty 1 year earlier (as part of the university's efforts to make its chemistry department one of the premier chemistry programs). With my declared interest in physical chemistry, the department placed Flory's name on my list of potential research advisors. At the time I knew very little about Professor Flory and the chemistry of polymers. In anticipation of meeting him, I set out to learn about him and his discipline by reading portions of his 1953 classic *Principles of Polymer Chemistry*. I found it fascinating.

So I met with Flory in his office in early 1963. Two small facets of that meeting, in hindsight, said something to me about Paul Flory. First was his modesty. In telling me about the research in his group, he referred to the "theta-temperature," a property of polymer solutions, as I later realized, that is commonly referred to as the "Flory temperature." But, he always called it "theta temperature." Also I recall during my first meeting with Flory that I was concerned that the department's deadline for submitting the name of my choice for graduate research advisor was coming up soon. Flory told me not to rush my decision, that I need not worry about the deadline. Flory's comforting reassurance, although a minor matter, turned out to be a preview of his protective nature which I experienced over the years.

Having found polymer chemistry very interesting and having found Paul Flory with the qualities I was looking for in a good mentor, I selected him to be my graduate advisor and joined his group. At that time it consisted of four postdoctoral fellows. Flory and his Dutch postdoc Agienus Vrij were then formulating a statistical mechanical model to describe the thermodynamic properties of *n*-alkanes and their solutions. Vrij's appointment ended just after I started, so Flory suggested that I pick up where Agienus was leaving off. This model required experimental densities, thermal expansion coefficients, and compressibilities of alkane liquids in order to predict the properties of their solutions—thus its designation as an equation-of-state theory. My first task was to comb the literature for experimental P–V–T data on alkane liquids and then put the data into forms that could be applied directly to test the new theory. I was very deliberate in my effort, meaning that I plodded along taking weeks to do what others would have accomplished in a fraction of that time. Flory was most patient with my snail-like pace. If he tried to pressure me to hurry, his efforts were too subtle for me to notice. I finally completed my undertaking, and Flory presented the new model at an American Chemical Society (ACS) meeting in January, 1964. Shortly thereafter he submitted for publication two manuscripts that included my compilation of literature data and their application to binary solutions of *n*-alkanes.

Research in Flory's Group

After I joined the Flory group, more graduate students and postdoctoral fellows came on board and its size grew to 10–12. Occasionally an undergraduate Stanford chemistry student added to that count. The photograph above (Figure 15.1) shows the group in the late summer of 1964. Flory helped set the tone for the group. The camaraderie was strong. We were mutually supportive and we had a good time together.

Flory research group, Summer 1964: Flory, Jim Mark (postdoc = pd), Bob Jernigan (graduate student = gs), Bill Gorth (gs), David Brant (pd), Akihiro Abe (pd), Steve Fisk (gs), Bruce Eichinger (gs), and Tony Semlyn (pd). Not pictured: Bob Orwoll (gs). Others who joined the group shortly after this picture was taken: Al Tonelli (gs), Al Williams (gs), Graham Blake (pd), and Wilmer Miller (sabbatical faculty).

Almost all of my graduate research centered on measuring equation-of-state properties of alkane liquids. Flory proved to be a wonderful mentor. After pointing me in the right direction at the onset, he watched my progress with great interest while giving me much independence. I and, I believe, the other members of our group found him positive and very encouraging as we conducted our studies.

The Flory group had a tradition of drinking tea at 3:30 each afternoon. The practice originated with an English postdoc before I joined the group and was still going strong when I left in 1966. The newest graduate student had the responsibility of boiling water, steeping the tea leaves, and announcing when the brew was complete. Graduate students, the occasional undergraduate, postdocs, Flory's secretary, and Flory would come together informally for 20–45 min in the lab. The usual discussion topic was chemical in nature, but not always. We found tea time to be ideal for reporting a new result—or problem—in the lab or telling of an interesting new finding from the literature. A nearby chalk board was frequently put into play. These unscripted, informal meetings over tea proved to be wonderful learning times. Flory apparently valued these daily sessions. He rarely missed them. He surely recognized their teaching value; and they also helped him keep abreast of happenings in the lab. They certainly added to the group's cooperative spirit.

Flory's research group occupied half the second floor of the Stauffer II Laboratory building at Stanford. The research group of inorganic chemist Henry Taube had the other half. The offices for Flory and Taube were part of a three-room suite across a hall from the laboratories. Access to the suite was through the middle room which was shared by Taube's secretary and Flory's secretary. Flory's office was on the left; Taube's to the right. Taube and Flory, both fine individuals, were respectful neighbors to one another. Their research groups interacted well. Remarkably, these two faculty with neighboring offices both won the Nobel Prize, Flory in 1974 and Taube 9 years later. Each was the sole chemistry laureate in his year. (I am most confident that neither lobbied beforehand for the prize, as some laureates reportedly have done. It would have been far out of character for either of these Stanford chemists.)

Flory had a strong interest in international human rights, particularly the mistreatment of scientists in the Soviet Union during the Cold War. As a graduate student, I was aware that Flory wrote letters and used other means to support Russian dissidents including some prominent individuals such as, I believe, Andrei Sakharov. However, I did not hear him speak much about these activities then. Now I regret that we did not ask him for more details.

However, later on I did get to see first-hand his efforts to help a young Soviet physicist named Yuri. This took place in 1981, near the end of my sabbatical at Stanford. (See Chapter 17, p. 169.) At that time Soviet authorities were blocking Yuri's efforts to leave the Soviet Union. He was being investigated because of unauthorized travel throughout Siberia. It turned out that Yuri had been surreptitiously delivering mail, medical supplies, and so on to Soviet intelligentsia who had been banished to the Soviet hinterlands for their political activities. Flory became aware of Yuri's precarious situation. So he invited Yuri to join his group, even though Yuri had no background in polymer science. The invitation by Flory, made especially significant by Flory's prestigious Nobel prize, must have put the Soviet authorities in an awkward position. The outcome: Yuri received permission to leave Russia. He got out and came to Stanford.

After Graduate School

I completed my experimental work by mid-1966 and defended my dissertation that October. Six weeks later I left Stanford for a postdoctoral position. Months earlier I had told Flory that I hoped to do a postdoc. He responded with the suggestion that I consider Walter Stockmayer, a widely respected physical polymer chemist at Dartmouth College. When I told Flory that that sounded wonderful, he offered to contact Stockmayer on my behalf. He did and Stockmayer responded by inviting me to join his group. The next 2 years at Dartmouth were wonderful. I could not have had finer mentors than Flory and Stockmayer. I remain most grateful to Paul Flory for connecting me with "Stocky."

As a graduate student, I had had few interactions with polymer chemists outside the Flory group. For instance, I had not attended American Chemical or American Physical Society meetings. In a sense, I had been sheltered. However, as a postdoc I began attending such meetings and met polymer chemists from outside the Flory sphere. On these occasions it was not unusual to be asked what it was like to work under the famous Paul Flory. I responded enthusiastically that it was wonderful. I pointed out that he was a caring, patient person, concerned about the well-being of us students, and protective of us, so much so, in fact, that he was a grandfather figure for me. Some of my questioners reacted with surprise. They said that my description did not match their impression of Flory. They told of witnessing him at meetings as a harsh and intimidating critic. This was not the person that I knew. But later I was present when Flory conducted himself as my questioners described. That is, I saw him respond to questions with little patience, very different from the kind of responses he gave to the questions we raised as graduate students in his lab.

One example took place at an ACS meeting in Philadelphia. Flory and I were talking in the hallway outside the meeting rooms, when a man approached and somewhat timidly addressed Flory. He was young, maybe a junior faculty member or a postdoctoral fellow. After superficially introducing himself, he began trying to explain a seemingly errant data point on a graph in a recently published paper that was a point of contention between his group and Flory. Shortly into the conversation, Flory apparently realized who the man was. Flory disagreed with the young man's assertions with anger and annoyance in his voice. The exchange was brief. The young man soon turned and walked away, looking disappointed and deflated. Then Flory turned to me and said, "Oh, no! I've done it again," acknowledging that he had verbally mistreated the young man and regretted doing so. It was as if he had been unable to help himself.

Sabbatical Year at Stanford

I maintained contact with Paul Flory after my graduation from Stanford. I returned to his lab during a sabbatical leave from my institution (William and Mary) for the 1980–1981 academic year. By then Flory had reduced his presence at Stanford to 3 days per week while working 2 days at IBM San Jose Laboratory in nearby San Jose. His research group at Stanford had declined in number to 4–6, all of whom were either postdoctorals or faculty

FIGURE 15.15
Walking on the Flory mountaintop property, spring 1981. Burak Erman, Burak's son Batu. Flory, Milenko Plavic, Wu DaCheng.

on sabbatical leave. I got to know Flory better that year. My office was almost next door to Flory's. Once in a while he asked if he might join me for lunch, as we usually both brought brown bags and ate at our desks. He seemed to enjoy telling stories from the past, even going back to his days at DuPont in the 1930s with Wallace Carothers.

Paul and Emily Flory enjoyed the outdoors, particularly hiking in the mountains. Probably for that reason, around 1965 they purchased a remote mountaintop property above the California coast overlooking the Big Sur area, a 2–3 h drive from Palo Alto. They had a house built on the land. It was accessible from California Highway #1 by an unpaved road, winding and narrow, leading a couple of miles back and up into the Santa Lucia mountains. The view from the house down to the Pacific was spectacular. In fact the viewing angle was steep enough so that one could see into the water, where, according to Flory, pods of migrating whales were sometimes visible from the front window. (The Florys kept a telescope mounted in that window.)

During my sabbatical year the Florys invited the research group to the mountain home for an afternoon in the spring of 1981 (Figure 15.15). Flory led some of us on an extended walk around the property and down into a beautiful redwood grove on the back of the mountain. He obviously loved this property. It is easy to imagine him finding it to be a peaceful getaway where he could relax and work without interruption.

Final Meeting with Paul Flory

My last meeting with Paul Flory took place at a lunch in June 1985. It followed a festive celebration of Flory's 75th birthday that included a symposium in his honor in the Chemistry Department at Stanford and an open house at the

FIGURE 15.16
Open house at the Florys', June 1985, on the occasion of Flory's 75th birthday. Pictured here: Akihiro Abe, R.W. Brotzmann, Bruce Eichinger, Matthias Ballauff, Walter Stockmayer, Paul Flory.

Flory home in Portola Valley (Figure 15.16). A large number of Flory's former students and research associates attended in addition to prominent scientists from outside the Flory family. I had not met some of these former associates before, but only knew of them as coauthors on Flory's papers. It was fun to meet them. On the day following the activities, Flory took Bob Jernigan (who had been a graduate student with me and was a close friend) and me to lunch at the Stanford Faculty Club. It was a special treat to be together with Bob and Flory. We had a delightful conversation. On that occasion Flory suggested that I spend the next summer with him to examine nonpolymeric liquid crystals, continuing work that we had started during my sabbatical. I told him that I would very much like to do that.

Three months later, on September 8, Paul Flory suffered a fatal heart attack at his mountaintop retreat, the peaceful place where we had visited the Florys 4 years earlier.

Agienus Vrij

I was educated in the Dutch School of Colloid Science of Kruyt (1916) and Overbeek (1946), his successor. My university study started in 1949 in Utrecht where I obtained my first degree (so-called Candidaats Examen) in chemistry/physics with mathematics in 1953. We obtained an extended education in physical chemistry in the "European sense," (e.g., the book of Rutgers[83], with a Foreword by Debye).

Hereafter I chose the Laboratory of Overbeek: The van't Hoff Laboratory of Physical and Colloid Chemistry. Overbeek had a broad interest in the physical chemistry of colloids and surfaces including polymers (polyelectrolytes). Charge stabilization and van der Waals-London destabilization was the major theme (the DLVO-theory of colloid stability[84]). But also other charged particles as ionic micelles were investigated. Subjects were electrochemistry, colloid science, polymer science, about 60 lecture hours altogether. Next to this Overbeek gave his "Capita Selecta" on light scattering, sedimentation, micelles, and so on, derived from his own research. As further, smaller studies we had to take courses in another department, which was in my case theoretical physics.

My experimental work as a (say master) student was centered around light scattering with the Brice-Phoenix apparatus. My doctoral exam was in 1956. My thesis work was on the light scattering of charged particles in salt solutions[85] for which I had developed a new theory, which was supplemented with experiments on Na-poly methyl methacrylate in aqueous salt solutions.

After my thesis I went into military service for nearly 2 years after which I returned to the Overbeek laboratory. Soap lamellae (as in soap bubbles) as a model for interaction forces was the theme. Here I started research on light scattering of surfaces of such very thin lamellae; a new technique. In the mean time I searched for a postdoc position in the United States and was admitted in Stanford to the still small team of Professor Flory. There were four other postdocs at that time in the department: Brian Jackson (UK), Dave Brant, Vittorio Crezenzi (Italy), and Jim Mark.

The subject of my project was a theoretical one. Professor Flory explained in half a page his new model for the thermodynamic properties of linear, liquid alkanes. It was a sort of van der Waals theory but then for polymer segments with so-called external degrees of freedom (after Prigogine). Ultimately Flory wanted to come to a better theory of polymer solutions using what he called Equation-of-State properties as input for parameters he needed for his equations.

My task was to collect these properties for alkanes with, say, $n = 1–40$ from the literature. Unfortunately, such properties of a sufficient large scope were not present in chemical abstracts. Therefore, Flory suggested to measure densities and expansion coefficients myself. For this I needed glass instruments that could only be made (in my perspective) by a professional glass-blower. But in all of Stanford there was only one glass-blower who did not have time to help me. (Compare this with the situation in the Netherlands at that time, where each chemistry laboratory had at least one but usually more glass-blowers.) Thus this came to nothing and Flory gave me another subject: determine the melting point of polyethylene, as an extrapolation of the alkanes. This became succesful and a paper was published (1963). Further I myself took upon me the task to extend the theory to the mixtures of alkanes and derive the formulae for that situation. This was written in a report which I left with Flory. In the spring of 1963 I found by accident a rich source in the chemical engineering literature, but it came too late then to finish the job.

After me, Orwoll was put on the job as a doctoral thesis student. The work was finished and published in two papers in 1964, which became rather succesful as the Flory–Orwoll–Vrij (FOV) theory. Up to now references are made to it, about 800. The theory was later extended by Flory and coworkers to a more general class of systems.

Returning to Utrecht I became an associate professor in 1966 and a full professor in 1968. In my research program several topics were connected with polymer science. A theoretical project with Overbeek was on the stabilization of colloid particles by adsorbed or chemically attached polymer chains. The idea was to prevent the effect of attractive van der Waals-London forces at short distances by covering the particles with a "loose" layer of polymer chains. A theoretical study was started by thesis student Hesselink in 1966. The particles are repelled by entropic repulsion of the polymer chains which have lesser space to develop configurations between nearing particle surfaces. Further Flory–Krigbaum interaction forces operate between overlapping chain clouds. The interactions can be manipulated by varying temperature and solvent quality.[86]

In 1971–1972 I was a visiting professor at The Massachusetts Institute of Technology, Department of Chemical Engineering. In 1990 I was honored with the Koninklijke/Shell prize for my whole oeuvre. In 1993 I took early retirement. The number of my PhD students was 22. The number of our publications was around 125.

Robert L. Jernigan

Paul Flory—A Strong and Personable Scientist

From the beginning as a graduate student it was clear that Flory was a serious researcher—he would immediately suggest what he wanted and push you ahead, as a fresh grad student, to start on a project—he was a man of action, and could be impatient if results were not forthcoming. The lessons that I learned from Flory about how to carry out research were certainly as significant as the individual research findings.

Personal interactions with Flory were intense, but never intimidating. At one point we were working out the higher moment expressions for different chain models—freely jointed chains, freely rotating chains, worm-like chains and rotational isomeric chains. At the beginning Flory would hand me a sheaf of papers that included long derivations of some of these expressions. He would ask me to check them, and to use these examples to push ahead to derive further results. His sense of urgency was always present and was contagious and led to some excitement upon completing results. But, being a man of action he always would do what he needed to get a new project started. The close personal interactions between Flory and his students were critical for making these efforts succeed, and he would always quickly draw the students into a new project.

I also recall times at the weekly group meetings held in the gazebo between the Stauffer I and II buildings when he would pose questions to the student presenting that were usually simple, insightful, and pushing the student to expand his thinking. His examples of learning to ask the right questions were wonderful. He could also express extreme displeasure whenever there had been little progress on a project. I had a roommate who began working with Flory but could not keep up, and in one of these sessions his poor progress became embarrassingly apparent, and on the spot Flory told him to leave. This level of abruptness and impatience was always well justified, and you always knew that Flory was fair in his decisions, but he was nonetheless a formidable adversary, especially if you were a student. He always made his expectations clear, and when you met or exceeded them you were rewarded with his personal warmth and large grin. One of the ways this would come out was in his blunt and outspoken opinions of other scientists.

Much of the pleasure that I had in working with Flory was from the pleasure of intense intellectual interactions. In the course of computing various averages, we would pass pages of equations back and forth, revising and correcting each another. It was extremely pleasurable, and at an intensity level usually of nearly one exchange per day. It was a pleasure not to be sticking to a schedule while doing this, but just to meet as quickly and as often as possible. This level of intense collaboration was really one of the greatest pleasures in the Flory laboratory, which I continue experiencing in interacting with my own students.

He believed in publishing every piece of research, even minor pieces. However, that being said, he would confide that he had a drawer full of unpublished manuscripts that had not successfully passed through the review process, and many of which he viewed to be especially important papers.

Primitive Computing

I recall early on that he was just beginning to use computers and had been doing this with a graduate student in computer science, but he wanted faster results and more direct input into it, so, as a new student, he suggested that I take up the project, and he handed me a small deck of computer cards and had me meet with the computer science grad student who had been working on this. I was to learn Algol since this was the language this program had been written in—revolutionary, because it could have more than one statement per card since it had punctuation between statements. I recall many days and nights spent sitting in front of card punching machines in the wood-framed Eichler building where the Burroughs B-5000 computer was housed. The fear there was that you might drop the boxes of cards and lose everything when they were shuffled, though we could put a number on the cards so they could be reassembled after such a mishap, but this was not so easy if you were continually changing your program, as we were always doing.

I also recall the day when Flory bought for the lab a new Marchant desktop calculator that was revolutionary because it had one number that could be stored and called up as needed. We were continually using this to check the output from the "large" computers since we were not always fully confident of our self-taught programming skills, or I suppose of the computers.

One lesson I learned from this is that there is no more important incentive for a student to learn something new than to have an urgent need for it in his research project, and that even difficult things can be learned quickly under these circumstances.

Views on Science

Flory's views on the importance of models and computations being held to match or reproduce experimental data was a strong lesson. He always expressed strong distaste for computations and simulations that were not tightly tethered to experimental data.

Personal Interactions

When a visitor was coming he would always offer a fully candid opinion about the person, which was sometimes astonishing to a young student. Part of this was really helpful because it provided guidance for how much effort we should put into the discussions, and how seriously we should discuss our own research with the visitor. As a student, it was a bit surprising to hear his honest and strong opinions of others. Balancing this, he also expressed strongly favorable opinions about many others. This was particularly useful when it came to looking for postdoc mentors; when asked about individuals Flory was always blunt, in a nearly black and white way; it seemed scientists were either wonderful or terrible. From my experience I found that his opinions were always insightful into both the science and the personalities. He had an uncanny and nearly instantaneous ability to size up scientists and to decide whether they were doing interesting research or not.

He had a particularly strong interest in scientists from the Eastern Bloc. He always expressed strongly favorable opinions of the Russians, Mikhael Volkenstein and Oleg Ptitsyn, for example. These opinions had a strong influence on me personally. Volkenstein's new book on *"Configurational Statistics of Polymeric Chains"* was the focus of some initial studies when I was a student. Later, when the Soviet Union was beginning to disintegrate, I was able to invite Oleg Ptitsyn to my lab at the NIH (National Institutes of Health) for a multiyear visit. After receiving the Nobel Prize, Flory engaged in important efforts to help Polish scientists. He considered support of human rights issues to be a personal responsibility.

David Brant

Personal History

I was born in Kingsport, Tennessee on September 20, 1936. My parents moved to Kingsport from Ithaca, New York following my father's completion of a PhD in organic chemistry at Cornell University in 1935. My father was employed in Kingsport by Tennessee Eastman, now Eastman Chemical Company. Tennessee Eastman was founded in 1920 to exploit the forests of the Appalachian region to produce essential Kodak starting materials, methanol, ethanol, acetone, acetic acid, acetic anhydride, and so on, from the destructive distillation of wood. Later it developed cellulosic polymers, polyesters, and, during the Second World War, became a major producer of military explosives and was involved in the management in the Y-12 plant at Oak Ridge. Late in 1943 the family began a series of moves related to my father's employment that found me attending school in Charlottesville, VA (elementary), Auburn, ME (middle), and Arlington Heights, IL (high).

For undergraduate studies I attended Yale with intentions from the beginning to pursue a chemistry major. My PChem teacher, Jon Singer, who subsequently became famous after moving to UCSD (University of California, San Diego) for proposing the 1972 Singer–Nicholson fluid mosaic model of the biological cell membrane, turned me on to physical chemistry. Undergraduate research in Singer's lab using density gradient ultracentrifugation to separate biological macromolecules and grad courses at Yale in general biochemistry (with Joseph Fruton) and protein biochemistry (with Julian Sturtevant) helped point me toward biophysical chemistry. A grad course in chemical thermodynamics (with Lars Onsager as occasional lecturer) cemented my interest in that topic, and I consider J. W. Gibbs a distant professional ancestor.

I left Yale in June 1958 with a National Science Foundation (NSF) Graduate Research Fellowship in hand ($1600/year) and went immediately to Madison, Wisconsin for graduate school. That summer I took an intensive course in Russian and met my wife-to-be, Marian Freed, a Madison native who was a senior in history at UWM. With the NSF fellowship I had my choice of research groups in the UWM Chemistry Department. Of the three groups that I considered, that run by Bob Alberty was by far the most dynamic and appealing, although the focus of the group was on enzyme reaction mechanisms and my interests were more toward biomacromolecular structural issues.

I remember clearly rejecting a project proposed by ultracentrifugation pioneer, Jack Williams, who wanted me to investigate his idea that some proteins were made up of dissociable subunits. I associated this proposal with an early concept, roundly denounced by Sturtevant, that all proteins were aggregates of smaller units of about 5000 amu. Williams was on the right track, but I have no regrets about my inadequately researched decision. I also turned down a chance to join the group of polymer rheologist, John Ferry. Ferry was no longer working on proteins at the time, and I had absolutely no interest in his shake, rattle, and roll approach to polymer science. Since then I have overcome my prejudices on that front as well.

In the Alberty group I rubbed shoulders with such important subsequent contributors to biophysical chemistry as Gordon Hammes and Victor Bloomfield. These colleagues wisely advised me to abandon my plan to minor in biochemistry and to do the required minor in physics instead. I learned more of value from the junior–senior level physics courses than from any others I took at Madison. Probably most important for my future career was contact with postdoc Leonard Peller, who had done his PhD at Princeton with Walter Kauzmann. As a student in Sturtevant's class at Yale I had been given Peller's thesis to read. It developed a statistical mechanical theory of the helix-coil transition in synthetic polypeptides and was very influential in orienting my interest toward the conformational properties of biomacromolecules. Leonard was a great mentor, and a devoted enthusiast of the work of Paul Flory. Just as I was finishing at Wisconsin, Paul was moving from Mellon Institute to Stanford. I was lucky to get into his group as a postdoc soon after he arrived there. This opportunity tested my antagonism toward the West Coast, and I soon learned the animus was misplaced. Acting on an idea suggested by Paul, I applied for and got an NIH postdoctoral fellowship ($6000/year) to pursue Flory's model for muscle action based on the polypeptide helix-coil transition in crosslinked polypeptide chains before we left Madison for Stanford in the summer of 1962.

At Stanford

I began work in the Flory lab studying the properties of concentrated solutions of helical polypeptides, where they form liquid crystalline phases, as a precursor to preparing crosslinked helical bundles as a muscle fiber model. Paul had written a 1956 paper in the *Proceedings of the Royal Society* (PRS) that developed a lattice theory of the lyotropic phase transition observed in solutions of rodlike polymers, a topic addressed initially by Onsager in 1949 and one that Paul pursued actively with Aki Abe and others during my time at Stanford and afterwards. Solutions of rodlike α-helical synthetic polypeptides had been observed to convert from isotropic to anisotropic behavior at sufficiently high polymer concentration, with a biphasic region in a narrow range of concentration. Although I had taken a course from John Ferry based on *Principles of Polymer Chemistry*, I needed to dig deeply into the bible to come to grips with the *PRS* paper. This project led me to work through the entire *Principles* book during my first few months at Stanford. Our colleagues Agienus Vrij, visiting from Utrecht, and Vittorio Crescenzi, from Naples, were embarked on the same course of study and proved to be helpful interlocutors. Studying the *PRS* paper also induced me to try to reproduce the phase diagrams Paul had published. This was the impetus for me to learn to program the available Burroughs computer, since I had no desire to do the calculations by hand as Paul had done. I learned the programming language of the day, BALGOL (the Burroughs strain of ALGOL) and was able to solve the necessary equations. This exercise stood me in very good stead for what was ahead. The experimental side of this project, using a falling ball viscometer to determine the phase diagram for the specific polypeptides we had purchased, did not get very far before I was off in another direction. We did determine that it was easy to locate the phase boundaries in a temperature–composition phase diagram, given the much lower viscosity of the anisotropic phase.

There was in the lab a new grad student who had been put to work on applications to polypeptides of the statistical mechanical theory of polymer configuration that was occupying much of Paul's attention at the time. The student left, and Paul asked me to switch my attention to the solution behavior of random coil polypeptides. I did not get back to studying the liquid crystalline phases in solutions of rodlike polymers until near the end of my career at UC Irvine. This was a very fortuitous change of direction for me. I immediately became involved in computer modeling of polymer chain configuration involving very early applications of molecular mechanics to polymer conformation and statistical mechanical analysis of polymer configuration in solution. At that time single molecule experiments were unheard of, and measurements on polymers in solution and their interpretation necessarily involved studies of the ensemble-averaged properties of flexible chains. I determined the number-averaged molecular weights and osmotic second virial coefficients of three synthetic peptide homopolymers in appropriate solvents using an automated Mechrolab osmometer. Chain dimensions were deduced from the intrinsic viscosities using Ubbelohde viscometers and the Flory–Fox equation. Theta conditions were not accessible, and in order to get estimates of the unperturbed dimensions for comparisons with the theory we used the Flory–Orofino theory connecting the expansion factor to the second virial coefficient. We got consistent values of the characteristic ratio for all three polymers and quite satisfactory agreement between experiment and theory. This work formed much of the basis for Chapter VII of *Statistical Mechanics of Chain Molecules*. About half way through my time at Stanford another of Leonard Peller's recruits to the Flory lab, Wilmer Miller from the University of Iowa, arrived for a sabbatical visit. I had taken over Wilmer's desk in the Alberty lab, since he was leaving Wisconsin just as I arrived, and at Stanford we began a fruitful collaboration focused on our mutual interest in polypeptides.

Paul had acquired from France a new SOFICA light scattering instrument, at the time the best commercial instrument available. Bob Chiang from Chemstrand visited the lab with some linear polyethylene fractions that he wanted to characterize using the instrument. This was a great opportunity for me to learn something about light scattering, a tool that I subsequently used extensively after leaving Stanford. Bob's task required working with α-chloronaphthlene as solvent at a temperature of 140 C. Under these circumstances we needed to replace benzene, the usual index matching fluid used in the SOFICA, with silicone oil. Overall, this proved to be a very messy business including clarification of the solutions at elevated temperature and cleaning the oily cells, but I became familiar with the light scattering technique and used it to advantage later. In this context I also became aware of the pitfalls that may attend light scattering measurements in mixed solvent systems.

I remained at Stanford for 3 years involved in work that positioned me for a faculty position in a research-intensive institution. Again Leonard Peller, who had by this time moved to UCSF (University of California, San Francisco), played an important role in determining my future direction. As I was beginning to interview for jobs in early 1965, Leonard mentioned that a friend he had known at Princeton, Sherry Rowland, had accepted the job of starting the Chemistry Department at the new UC Irvine campus, slated to open in the fall of that year. I got in touch with Rowland and was soon in Irvine for an interview. I accepted the job offer that came shortly thereafter, because I sensed the opportunity to have an immediate, and perhaps important, influence on what developed. Of course there were distinct disadvantages to going to a brand new campus, for example, few if any graduate students. I was lucky enough to land an NIH R01 grant even before arriving at UCI (try that now!), so I was able to attract one of the few new grad students and to support a postdoc right away. I had decided in the process of writing the NIH proposal to switch my focus from proteins and polypeptides to polysaccharides. The large and productive groups run by Harold Scheraga at Cornell, Paul Doty at Harvard, and Murray Goodman and Bruno Zimm at UCSD seemed to me to present a level of activity in polypeptides that would be difficult to compete with from a new lab in a new institution with an uncertain supply of students. This proved to be a good decision. I was able to devote my research effort during a 40-year career at UCI to the study of carbohydrate polymers, as new and interesting physical and biological attributes kept emerging almost every year. Since retiring from UCI in 2005, I have been a program officer in the NSF Division of Materials Research principally involved in starting and managing the biomaterials program.

Reflections

I think those of us in the Flory lab in the early 1960s had perhaps a unique opportunity to engage with Paul in a personal way. He had shed the management responsibilities associated with his job at Mellon, and administrative duties at Stanford had not yet intruded on his time. He was thus able to both relax and direct much of his attention to the research topics that motivated him. He came into the lab every day for a group tea party, preparation of which I think he delegated to one of the Englishmen in our midst, Brian Jackson and Tony Semlyen among them. I was most impressed by the volume of new ideas Paul brought to the lab each morning after an evening of thinking and writing. We, especially Bob Jernigan, Bruce Eichinger, and Bob Orwoll, were often asked to check some new equations he had come up with overnight. A number of very successful academic and scientific careers were launched from that environment, owing much to the standards of diligence and excellence he set for us. Wilma Olson, Paul Schimmel, Wayne Mattice, and Ken Dill who arrived in the Flory lab soon after I left, are among those whose careers display clearly the significant influences Paul had in the area of biopolymers.

Paul and Emily periodically invited the group for get-togethers at their house on a ridge overlooking the Stanford campus with views across the Bay to the East Bay hills. My wife and I were asked on one occasion to occupy the house for a couple of weeks while they were away. Our duties were to keep the grass mowed, water the plants, and perform periodic chemical incantations over the swimming pool. Paul worked very hard, but not to the exclusion of some breaks for recuperation. I remember well, with some residual envy, a canoe trip Paul and his son Jack took to Glen Canyon before the dam was built and another to the remote John Day River in Oregon. Paul was an outspoken fan of David Brower, the Sierra Club, and the wilderness movement. Before he acquired his house above Big Sur, Jim Mark, and I and our wives had hiked right past the house site on a backpacking trip we took to the Ventana Wilderness, without knowing that Paul already had his eye on that spectacular place. Later Paul became a passionate advocate for the political rights of East European scientists.

Bruce Eichinger

Reminiscences of P.J. Flory

As an undergraduate majoring in chemistry at the University of Minnesota I had the good fortune to conduct research with the physical chemist, Stephen Prager, beginning in the summer after my junior year, 1962. I wrote a program for evaluating the diffusion coefficient for fluids migrating through a bed of two-dimensional discs. The power-house computer at the University of Minnesota (UofM) was a CDC 1604 with 32 K of 48-bit words, which was the first transistorized computer designed by Seymour Cray, no less. (The Wikipedia article about the CDC1604 will provide more details for those interested in the early days of transistorized computing.) To do the calculations I had to write machine code to pack data into the fixed word size to make efficient use of memory. More than once my inept memory management skills invaded the operating system, which required the computer center staff to reboot the machine. To control the program I was allowed to listen to the electronic noise of the processor and manually throw a sense-switch when the placement of the next disc was taking too long. Computer time cost hundreds of dollars an hour, so it was important to be careful with the time. Were those the good old days?

When it came time in the autumn of 1962 to apply to schools for graduate work, Prager advised that I apply to X, Y, and Z, but said that I should go to Stanford to study with Flory. I did as I was told, and happily was accepted by Stanford. I got a head start on polymer chemistry during my last year at the UofM by taking a one-quarter course offered by Professor Prager.

The incoming chemistry graduate students at Stanford in 1963 were a tight-knit group. Amongst our number were Barry Sharpless and Tom Meyer. After passing the dreaded prelim exams on the second try, I got to work in Flory's lab with graduate students Bob Orwoll and Bob Jernigan. At the time the postdocs were Jim Mark, Dave Brant, and Bill Leonard. First year graduate student Steve Fisk also joined Flory's group that year. Flory put me to work on polymer solution thermodynamics, which wound up being the title of my thesis in 1967.

I inherited Bill Leonard's vacuum line apparatus and started my research with measurements of the sorption of benzene vapor by natural rubber. Bill's equipment was modeled on that developed by Stephen Prager in his PhD work with F. Long at Cornell, which is where the Flory connection was established! In the 1960s, Flory had two major focuses: thermodynamics of solutions and rotational isomeric state theory. Those of us working on thermodynamics, including Bob Orwoll, developed techniques to measure thermal expansion coefficients and volume changes on mixing that were remarkably accurate at the time. Throughout my graduate career the lab was a-hum with stirring motors and flashing lights that controlled our constant temperature baths.

Perhaps Flory was being cautious in having me work on natural rubber and benzene. This system had been studied previously by Gee and Treloar in their seminal work published in 1942, very shortly after Flory and Huggins independently devised polymer solution theory. Gee and Treloar showed that the Flory–Huggins polymer solution theory gave a perfect account of the thermodynamics of this system, with the interaction parameter χ having a value of about 0.4, independent of concentration. It has turned out that rubber-benzene is one of the very few systems with a nearly constant value of χ—what luck! In any event, the Gee–Treloar experimental work was definitive proof that the entropy of polymer solutions is very different from that of mixtures of small molecules.

Flory's lab had a very dynamic work ethic, exemplified by Flory himself. His standard practice was to work on Saturday mornings, and many of us saw this as an appropriate time to be in the lab as well. I usually came into the lab in the evenings to put in a few hours at the vacuum line degassing polymers or taking a measurement for the thermal expansion coefficient of polyisobutylene (PIB) (which Flory asked me to repeat twice!).

In the middle 1960s several postdocs and graduate students migrated through the Flory lab. In addition to the postdocs Bill Leonard, Jim Mark, and Dave Brant, the lab hosted Hartwig Höcker, Paul Schimmel, Wilmer Miller, Akihiro Abe, Tony Semlyen, Fumiyuki Hamada, and Graham Blake, not necessarily in that order. The graduate students who joined a year or so after I started were Al Tonelli and Allen Williams, both of whom worked on rotational isomeric state (RIS) theory. Bill Gorth and James Ellenson, both undergraduates, also worked in the lab for about a year in that time frame. This was a distinguished group of colleagues, and I am grateful to all of them.

When I finished my dissertation in the spring of 1967, Professor Flory was kind enough to keep me on as a postdoc until the autumn when a postdoctoral appointment with Marshall Fixman at Yale would commence. During my last year at Stanford I advised Ellenson, who took up measurements on volume changes on mixing of normal hydrocarbons with polyisobutylene (PIB). This polymer was an absolute bear to handle, because all our measurements required careful degassing of the polymer, but gaseous diffusion coefficients in PIB are the lowest of any polymer known at the time. Ellenson and I both suffered long days on the vacuum line carefully controlling the temperature and pressure on the PIB samples to degas them prior to measurement. In any event, I was happy to have contributed to what I believe was the first theoretical paper showing that two polymers, in this case PIB and polyethylene (PE), are immiscible.

During this short postdoctoral time I worked on a theoretical problem in gelation that PJ asked me to look at. Flory was *the* pioneer in gelation in 1941 using ideas known as branching theory. Perhaps by 1967 it was becoming evident that the branching theory was in difficulty, as it leads to an exponential growth in the number of off-spring as the generations increase. His idea was to start with branching theory, for which the incipient network is approximated as a tree with no cycles, and treat cyclization as a problem of attrition in successive generations. A cycle that forms is equivalent to killing off one or more branches of the family tree. I produced a few innocuous statistical relations between the structural parameters, but was unable to generate anything useful. However, Flory's request was sufficient to get me hooked on elasticity and network structure, where I spent a large portion of my career.

Teatime in the Flory lab was an afternoon social event that persisted throughout my time in the group. We took turns boiling up a large beaker of water in which the tea was made for our afternoon ritual. This was a time for the several of us in the lab, including Professor Flory, to get together for some informal chats and discussion. I looked forward to tea, as Flory would sometimes tell us about some of his experiences, talk about current literature, and give us some insights into other polymer scientists. However, Flory was exceptionally modest in these social situations, and was a rather serious man, so there was not much laughter in the lab.

However, laughter did prevail one day. Flory's secretary—to the best of my recollection this English lady was named Heather—came into the lab for tea almost doubled over with laughter. We asked what was so funny, and she related, after some coaxing, that Flory could dictate a marvelous letter. The letter in question was to the U.S. Department of State, in which Flory defended his actions of a few months earlier. An unnamed Russian scientist had visited Stanford and subsequently complained to our State Department that he was not treated well by Professor Flory. As I remember, Heather's rendition of one of the sentences in the letter went something like this—Flory is writing—"I will be more than happy to entertain Russian scientists, but if you send another Russian bureaucrat I'll throw him out of my office too!" This has to be understood in the context of Flory's unfettered support for Russian scientists; he was not jingoistic, but he did not abide anyone who disagreed with him. The heated literature controversy over rubber elasticity that ran for several years between Flory, championing the Flory–Wall theory, and the physicists James and Guth, is further evidence of Flory's combative spirit in scientific matters. Hopefully others will comment on Flory's work in support of Russian scientists through the difficult 1960s and 1970s in the Soviet Union.

After a wonderful year with Marshall Fixman at Yale, I was appointed assistant professor in chemistry at the University of Washington commencing in the autumn of 1968. In starting up my research program I needed a laboratory sample of polydimethylsiloxane, so I called Professor Flory to ask what might be the best way to get a sample of the polymer. Flory gave me a telephone number for Art Bueche at General Electric (GE), who would likely be able to supply a sample. I knew of the Flory–Bueche connection, so I gave him a call. I did not know that I was calling the VP for Research at GE! In any event, Dr. Bueche was very gracious and I received a nice sample of high molecular weight polydimethyl siloxane (PDMS) a week later. This turned out to be a much better way to get a laboratory sample than my experience with polyethylene. I asked someone at Dow to send a sample of PE, and in a few weeks our stockroom wanted to know what to do with a pallet of four 50 lb bags of PE pellets!

Discussing Flory's contributions to chemistry is a daunting task—where does one start? Let me pick out just a few examples to illustrate his exceptional talent. The first of these was his insightful avoidance of hopeless intermolecular potentials to describe the interactions between segments of polymers. Flory cut through unrewarding details and used macroscopic thermodynamics to describe these interactions. This was the basis for his work on the second virial coefficient and the excluded volume expansion factor. Flory said that when he came up with the theory of the expansion factor while at Cornell, Peter Debye vigorously disagreed. Debye thought that the expansion factor should asymptote with increasing molecular weight, whereas Flory's theory has it go as a power of the molecular weight (MW). Flory reported that he went the other way when he saw Debye in the hall! Of course Flory was right, and the theory of the expansion factor, and in particular the observation that the theory implies that chain configurations in bulk amorphous polymers are unperturbed, was cited by the Nobel Committee.

As my appreciation for Flory's insights has grown over the years I have come to think of Flory as "The Man Who Understood Entropy." Underlying his mastery of entropy was a profound facility with all things statistical. Two papers in particular stand out as extremely ingenious and clever applications of statistical arguments to chemical problems. The earlier of these "Intramolecular reaction between neighboring substituents of vinyl polymers," *JACS*, 61, 1518–1521 (1939) and the second "Molecular size distribution in three dimensional polymers. I. Gelation," *JACS*, 63, 3083–3090 (1941) were produced while Flory was at the University of Cincinnati. The first involves a clever use of "matching" statistics combined with a dazzling mathematical analysis of the equations. However, his real genius was revealed in the papers on gelation. The argument that Flory conjured up, using what is now known as branching theory, reduced an apparently hopeless complex problem to amazingly simple terms. When the number of daughters in a generation selected at random equals or exceeds the number of mothers, the family tree will persist forever. This provides a simple algebraic relation that specifies the point at which the fertility index (the average number of daughters in a generation) is sufficient to guarantee survival of the family tree. That paper provided the basis for more than one career in polymer science. The theory was taken over by mathematicians, particularly Erdős and Rényi around 1960, who rarely, if ever, give credit to Flory's seminal work.

An effective mentor is one who instills a love for work and the desire to keep learning. Flory was a master at this. Most of his students and postdocs have emulated his work ethic and love for science—on my better days I am one of those. I am grateful to be a Flory student.

Alan E. Tonelli

My Flory Remembrance

My Personal Narrative

During 1962–1963, my junior year in chemical engineering at the University of Kansas, I was fortunate to be taking physical chemistry in the Chemistry Department with Professor Paul W. Gilles. My performance in his PChem class prompted him to ask me if I would like to try doing some research in his lab. I agreed, and this experience opened my eyes to and started me on the way to a long career in research. I quickly got caught up in the excitement of his graduate students, particularly as they discussed their latest findings among themselves and with Professor Gilles in their biweekly research group meetings. I suppose Professor Gilles recognized my excitement for research, because he took it upon himself to convert me to chemistry.

He made a strong case for going to graduate school and majoring in physical chemistry. After some cajoling, I agreed to take two additional semesters of organic chemistry, and one semester each of inorganic, statistical mechanics, and quantum chemistry during my senior year. I was only able to do this because of the cooperation extended to me by my chemical engineering professors who agreed that I need not attend, but only needed to complete the work assigned in their classes. Furthermore, Professor Gilles made sure I took the GRE (Graduate Record Examination), and, on my behalf applied for an NSF Cooperative Graduate Fellowship, which I was fortunate to obtain.

The NSF fellowship, coupled with only two job offers, as a Second Lieutenant with the Army Corps of Engineers and as a process engineer at a Sinclair refinery in Oklahoma, opened the doors to graduate school for me. Because NSF was paying my expenses for 2 years, I was admitted by virtually every graduate program I applied to. My recent conversion to chemistry prevented my selection of graduate programs on the basis of faculty quality or research emphases. Honestly, I chose Stanford, over the equally prestigious Ivy League universities that had accepted me, for its location in sunny California. Though admittedly ignorant, what a wonderful choice it was.

I began my graduate school experience at Stanford in the fall of 1964 by concentrating on passing the four exams in chemistry necessary to qualify for admission to the PhD program. Luckily, I was able to pass two at the beginning and the remaining two qualifying exams at the end of the first quarter. Though most graduate students joined research groups as research assistants very early on, I did not for two reasons. First, and most important, was the financial independence provided by my NSF fellowship. Second, was my indecisiveness in selecting a prospective faculty research mentor/supervisor. For the next year and a half I took several graduate courses in chemistry and undergraduate courses in mathematics, and also met my obligation as a teaching assistant by supervising a quantitative analysis lab.

My Personal Interactions with Professor Flory

However, in the spring of 1966, as the end of my NSF support approached, I bit the bullet by asking Professor Paul J. Flory if I could join his research group. My request was based on two factors: (1) Professor Flory was a physical chemist, and I was most comfortable with physical chemistry and (2) in my first year at Stanford I spoke to him about a thermodynamics class he was to teach the following quarter, and asked if he recommended it for me.

His direct and honest response was that my undergraduate chemical engineering degree likely made his course unnecessary. Instead he suggested I take another graduate course that was also being offered that quarter. This single interaction with him, and his down-to-earth, noncondescending manner, provided me with the courage sufficient to ask him if I could join his research group. My good fortune in joining the Flory group was immediately impressed upon me, when several of his graduate students insisted that I obtain a copy of *"Principles of Polymer Chemistry"* at the bookstore and begin religiously examining it. To my surprise and eventual delight, who should be the author of this classic polymer text, but my new faculty advisor.

As was the custom in the Flory group, the junior member was assigned the task of preparing afternoon tea, which I gladly did for more than a year. This consisted of boiling water in a large 5 L beaker placed atop a Bunsen burner, adding an ample portion of tea leaves, allowing them several minutes to steep, and then rapidly stirring to cause the saturated tea leaves to settle to the bottom. Professor Flory was then notified that tea was ready and would join us for tea and informal discussion. Flory tea time would invariably last for an hour or more, with discussion generally centered around our current research concerns or possibly a recent research development found in the literature. The tea time atmosphere was collegial, informative, and provided us with the opportunity to benefit from Professor Flory's vast experience in polymer science, and also to see how he approached research questions. We all enjoyed and profited greatly from Flory tea time.

At least once each month we would meet with Professor Flory over lunch and listen to, question, and discuss a research topic of current interest to our group, or one unearthed in the recent literature and presented by one of his students. These lunch time seminars provided additional contact with him, taught us how to organize, present, and evaluate research ideas, and afforded us with opportunities to think on our feet.

In 1967, Professor Flory asked his graduate students to carefully proof the manuscript version of his second book *"Statistical Mechanics of Chain Molecules."* If I remember correctly, we received academic credit for this task. We literally pored over his book, checking all equations and their derivations looking for errors. Fortunately we were aided by the expertise of Bob Jernigan, a Flory student senior to me, who had been intimately involved with developing the mathematics (matrix methods) for treating the statistical mechanical averaging of conformation-dependent polymer chain properties, such as $\langle r^2 \rangle$ and $\langle s^2 \rangle$. For me and my fellow Flory students, this was a unique and wonderful introduction and opportunity to learn about the conformations and statistical behaviors of polymer chains.

Parenthetically, in addition to his mathematical prowess and while a member of the Flory group, Bob Jernigan, with financing from his family, opened a laundromat. This entrepreneurial venture was rather short-lived, however, due to lack of profits, and more importantly, because of the large amounts of time and attention it took away from his research. Though at that time he may have lacked sufficient business acumen, Bob Jernigan went on to a long and highly successful research career at NIH.

Each year, Professor Flory and his wife Emily graciously hosted his research group, their significant others and families to a pool party and barbeque at their lovely home in the hills overlooking the Stanford campus. These parties were never occasions for "talking shop," but were instead strictly social gatherings, where we learned more about the Florys and in turn they about us. It is easy for graduate students to attribute "God-like" qualities to a scientist as eminent as Professor Flory. But on these social occasions, we could clearly see that he was wonderfully human, and this provided hope and inspiration to us all.

Without benefit from the early guidance and help of Professor Paul W. Gilles at the University of Kansas, it is more than doubtful I would have become a chemist and a researcher. In addition to the huge quantities of "dumb luck" that sent me to Stanford, the kindness Professor Flory showed by accepting me as a graduate student in his research group most certainly shaped my subsequent life.

Because of his reputation in polymer science circles, one soon to be validated by receiving the 1974 Nobel Prize in chemistry, the Director of Materials Research at AT&T Bell Labs, at that time Dr. Bruce Hannay, would visit the Flory lab each year. His visits were to learn what we were working on and to see if any Flory students were soon to graduate. On these visits, Dr. Hannay would extend an invitation and arrange a Bell Labs employment interview for eligible Flory students. In the spring of 1968 I interviewed at Bell Labs and a half dozen chemical companies. At Bell Labs I met and was hosted by Dr. Frank Bovey, of Bovey and Tiers fame, who had recently moved from 3M to head the Polymer Research Department. I was both familiar and highly impressed with Frank's ground-breaking studies of polymers using NMR, because I presented his early results at one of the Flory group lunch time seminars. It was a "no contest" decision to accept the Bell Labs offer, and there I spent 23 most enjoyable years conducting fundamental research in polymer science.

For the last 22 years I have been a professor of polymer science at NC-State University, where I try to treat, teach, and guide my students as I learned from Professor Paul J. Flory.

Thank you Paul!

My Research with Professor Flory

I begin by noting that in my research experience with Paul Flory, he never hovered over or "micro-managed" his graduate students. Instead, he suggested initial research topics/areas, welcomed reports of research progress or difficulties, and was surprisingly available for questions, even beyond those raised daily in our tea time discussions. We soon learned to use each other as initial sounding boards for our research concerns, and if no consensus was reached in our group discussions, we would raise the issue at the next tea time or directly approach Professor Flory in his office.

My PhD thesis research dealt with (I) The Configurational Statistics of Polylactic Acid and (II) Optical Anisotropy in Vinyl Polymers, as evident below in the papers I coauthored with Professor Flory. The first resulted from the conformational energy calculations performed on polylactic acid (PLA) by David Brant and Professor Flory just prior to my arrival. Using this conformational description, the dimensions ($\langle r^2 \rangle$) of PLA were predicted by them, and I set about to measure them using viscosity and light-scattering techniques. Memorable in this pursuit was the SOFICA light scattering instrument developed in France. Our SOFICA was the first in the United States, and its extensive manual was written solely in French. Needless to say that, because of my time with the SOFICA, I selected French as one of the two foreign languages in which I was required to demonstrate reading comprehension.

"Optical anisotropy of vinyl polymer chains. 1. Strain birefringence of polypropylene and polystyrene," Abe, Y., Tonelli, A.E., Flory, P.J., *Macromolecules*, 3(3), 294, 1970. DOI: 10.1021/ma60015a005.

"Optical anisotropy of vinyl polymer chains. 2. Depolarized scattering by polypropylene and polystyrene," Tonelli, A.E, Abe, Y., Flory, P.J., *Macromolecules*, 3(3), 303, 1970. DOI: 10.1021/ma60015a006.

"Configuration statistics of random poly(lacticacid) chains. I. Experimental results," Tonelli, A.E, Flory, P.J., *Macromolecules*, 2(3), 225, 1969. DOI: 10.1021/ma60009a002.

"Configurational statistics of random poly(lacticacid) chains. 2. Theory," Brant, D.A., Tonelli, A.E., Flory, P.J., *Macromolecules*, 2(3), 228, 1969. DOI: 10.1021/ma60009a003.

"Strain birefringence of polymer chains," Flory, P.J., Jernigan, R.L., Tonelli, A.E., *J. Chem. Phys.*, 48(8), 3822, 1968. DOI: 10.1063/1.1669690.

Because of solubility and refractive index issues, bromobenzene at 85°C was selected as the PLA solvent for light scattering. There were no lasers at the time, and the SOFICA sample cells were large (~20–30 mL). This made it frustratingly difficult to eliminate all dust and other particulates, which interfered with the inherent scattering from dilute PLA chains. Though humbling, this experience instilled in me a determination to succeed in finding experimental means to answer scientific questions, and, above all, the virtue of patience and hard work.

The second thrust of my thesis research was not nearly as time consuming and patience intensive as my foray into early light scattering. I used the matrix methods developed by Flory and Jernigan to calculate the dimensions, optical anisotropies, depolarization ratios, and strain birefringence expected of various vinyl polymers, using their Rotational Isomeric State models to average each of these properties statistically over their extraordinarily large numbers of conformations.

Both of these research thrusts have proved invaluable to me during the subsequent nearly 50 years of my career in polymer science, and remain so to this day.

Wilma King Olson

Memories of a Special Professor

Preliminaries

I had little, if any, idea of what I could do with an advanced degree in chemistry when I entered graduate school at Stanford University in 1967. Even though I had carried out an honors research project at the University of Delaware and held a summer internship at the Rohm and Haas Company Research Laboratories, I did not have the passion for research that often comes from these experiences. I applied to graduate school knowing what I could not and did not want to do with my undergraduate degree rather than what research area I wanted to pursue and/or with whom I wanted to study. My experience as a student teacher led me to believe that I did not have the depth of knowledge needed to start teaching chemistry in a secondary school, and I did not want an industrial position of the sort available at the time to a young woman with a bachelor's degree in chemistry, for example, chemical librarian. I had only a vague sense of what graduate school entailed and the directions in which it might lead me.

What inspired me to study chemistry were word problems. I had to choose a college major early in my senior year of high school as a stipulation of an early acceptance. Had I taken more physics at the time, I might have chosen a different major. What I came to love about both subjects was the application of mathematics to practical problems and the realization that the disciplines worked as one. Not surprisingly, I was at once fascinated by the mathematically based treatment of polymers developed in the Paul J. Flory laboratory.

Like many of my fellow graduate students from the East, I was intrigued by the chance to live in California. I had wanted to visit the Bay Area from the time, many years earlier, when my father went on an extended business trip to Redwood City. Although he described visiting the beautiful Stanford campus, it was William Mosher, chairman of the Chemistry Department at Delaware, who sparked the idea of my graduate study there. Professor Mosher led a seminar course in which he provided career guidance to the senior chemistry students. While his high praise of the chemistry program at Stanford may have been prejudiced by the fact that his younger brother

Harry was a member of the chemistry faculty, I had my own prejudices about applying to graduate schools. I had no interest in matriculating at a university that excluded women from its undergraduate colleges. Stanford met my criteria; the Ivy League then did not.

First Encounters with Paul Flory

What sold me about working with Professor Flory was the man as much as the research. Our first week on campus culminated in an evening welcoming party for the entering graduate students. I was surprised by the large turnout of faculty and students and thrilled by a delightful conversation with Professor and Mrs Flory. They introduced themselves to me knowing in advance who I was and where I came from and then sharing their past experiences in places familiar to me—their early careers with E. I. du Pont de Nemours and Company in Delaware, their life in northern New Jersey near the Standard Oil *Esso* Laboratories, and Mrs Flory's childhood outside of Philadelphia. None of the other professors were as warm and charming. Although the entire department had access to a document with the photos, names, and undergraduate schools of the entering graduate student class, only Professor Flory seemed to make use of this information. I learned by his example to study the corresponding information for subsequent classes, including those that I taught later in my career, so that I could greet new students in the same way that the Professor and Mrs Flory greeted me.

Our first week at Stanford included a series of qualifying examinations that we were required to pass before joining a research laboratory. With no background in instrumental analysis, I found myself among the group of entering students who had to make up academic deficiencies during the fall semester. I subsequently learned from two of my classmates, who were in the same situation, that we could attend the seminars of different research groups in order to find out about the opportunities that awaited us. We thought that our presence at these meetings went unnoticed until the day we first attended the Flory group seminar. Before we could make our usual quiet exit from the back of the room, Professor Flory strode up to us, expressing how glad he was to see us again, inviting us to bring our lunches to the next meeting, introducing us by name and where we came from to Al Williams and a few other members of the group, and asking Al to make copies of the schedule of upcoming speakers and topics for each of us. Needless to say, we returned several times that fall and once we successfully passed all of our qualifying examinations, scheduled appointments to meet one-on-one with Professor Flory.

These early meetings revealed even more about Professor Flory's character—in particular, how much more deeply he explained his work and how much more welcoming he was than some of the other professors. Professor Flory spent several hours going over the treatment of polymer chain statistics developed in his group and describing the applications of the methodology to various polymers. I was especially intrigued by his papers—with David Brant, Wilmer Miller, and Paul Schimmel—on polypeptide chains and their simplification of the polymer as a sequence of imaginary "virtual" bonds spanning the rigid peptide units. The work tied, albeit loosely, with my early love of word problems as well as with the measurements of enzyme activity that I had performed in Don Dennis' laboratory at Delaware and the nylon diffraction patterns that I had collected under Steve Yanai's guidance at Rohm and Haas. Professor Flory also told me about the graduate studies and lives of his two married daughters in stark contrast to another faculty member who discouraged my interest in his research group by asking whether I, then weighing less than 100 pounds, was capable of moving heavy pieces of equipment on a routine basis.

Studies of Polynucleotides

Not until I asked Professor Flory whether he would agree to serve as my research mentor did he show me a preprint that he had received from Heini Eisenberg of a study of a synthetic polynucleotide, polyriboadenylic acid, that Heini had just published with Gary Felsenfeld.[87] Heini autographed the cover of the preprint with a question about what Professor Flory might make of the unexpectedly high degree of chain extension that they had found. Professor Flory realized that the end-to-end length of the polynucleotide observed at higher temperature was comparable to that of a polypeptide and thought that it might be possible to interpret the system along lines similar to those used to study polypeptides.

Professor Flory had also recently contributed[88] to a two-volume book of papers presented in January 1967 at the International Conference on the Conformation of Biopolymers in Madras, India and received reprints of several papers written by G.N. Ramachandran, the meeting organizer and editor of the book. One of these papers described a preliminary study of the steric restrictions imposed on the conformation of a mononucleotide, the basic repeating unit of a nucleic acid or polynucleotide, and hinted of simplifications that might be introduced in a polymer model.[89] Professor Flory then left me on my own, departing for a 6-month sabbatical at the Weizmann Institute in Rehovot, Israel, where coincidentally Heini Eisenberg was a member of the Polymer Department that hosted Professor Flory's visit.

Professional Interactions

Like the other members of our group, my scientific interactions with Professor Flory came in spurts. He generally worked closely with one student at a time. My most valuable interactions with him were long one-on-one discussions of the drafts of documents that I was writing. Professor Flory set extremely high standards—expecting the research work to be of the very highest quality and the presentation to be thoughtful, logical, and crystal clear. We discussed the subtleties of wording and grammar and Professor Flory added some words to our papers for which I have never since found a use!* I sometimes still wonder whether he would approve of the text that I am struggling to write or the figures that I am composing. The illustrations included in my dissertation and in our joint papers were drawn from sketches that we submitted to an illustrator in the Medical School. I remember Professor Flory showing me how to check the accuracy of the two-dimensional energy contour maps associated with pairs of backbone torsion angles: the contour lines on the right and left and those on the top and bottom of the image should connect smoothly when the paper is rolled and the edges are brought together. I have subsequently used the trick in refereeing manuscripts.

Professor Flory held strong positions about many scientific issues. The three papers that we submitted for back-to-back publication in *Biopolymers* received uniformly positive reviews for scientific content and an added list of demands from one reviewer about the convention and nomenclature that we should use to describe the conformation of the polynucleotide chain. Whereas we assigned a value of zero to the fully extended, *trans* arrangements of the torsion angles formed by three successive chemical bonds, the referee insisted that we adopt the convention proposed for polypeptides by the IUPAC-IUB† Commission on Biochemical Nomenclature, in which the fully compact, *cis* arrangement would correspond to zero. Professor Flory had previously expressed his disagreement with the commission on this point and took the liberty of writing directly to G. N. Ramachandran, whom he suspected to be reviewer, about the strong rebuttal letter that he was sending to Murray Goodman, editor of the journal. I have often delighted in the points in his argument, one of which is copied below, and in his comments in the same letter about scientific commissions.‡

> For obvious logical reasons, and for pedagogical ones as well, the reference conformation should correspond to one written on a blackboard or appearing in a figure in a book. The angle $\varphi = 0$ should naturally correspond to the reference state. I defy anyone to draw an intelligible diagram of the all *cis* form of a polynucleotide chain, or even a polymethylene

* The following sentence—from Olson, W.K., Flory, P.J. Spatial configurations of polynucleotide chains. II. Conformational energies and the average dimensions of polyribonucleotides, *Biopolymers*, 11(1), 25–56—is a classic example of Professor Flory's inimitable vocabulary: "The analysis of the local conformations and of the spatial configuration of the chain as a whole, carried out on this basis, was nevertheless illuminating and served to adumbrate the principal interactions requiring more critical investigation." Adumbrate is one of the words I have yet to use.

† Acronym for the International Union of Pure and Applied Chemistry and the International Union of Biochemistry.

‡ Extracts from the letter written by Professor Paul J. Flory on September 24, 1971 to Professor Murray Goodman with carbon copies to Professor G. N. Ramachandran and Dr. Wilma K. Olson.

> chain having more than five bonds for that matter. Invariably the *trans* form is shown in diagrams. It ought to correspond to $\varphi = 0$.
>
> The view that recommendations of a commission should be adopted unquestioningly and universally is particularly to be deplored. When this kind of bureaucratic dictation prevails, it will be a sad day indeed for science. The recommendations of commissions deserve to be heard, but the obligation of the scientific community to adopt them is quite another matter. The language (and notation) of emerging branches of science should be determined not by official decree but by usage. Language evolves under the test of use and usefulness, not by decree. Thus, lexicographers codify language; they do not create it.

Professor Flory nurtured the careers of his students in numerous ways. For example, he gave me the opportunity to develop a project and then mentor one of the Stanford undergraduate students enrolled in a special honors chemistry laboratory course. He also asked my husband and me to assist with the local arrangements for a few of the many Flory laboratory visitors, some of whom later invited us to visit them and present our work in exotic locales. Professor Flory introduced me to the scientific jet set in 1969 when he sent me on a 1-day excursion to Los Angeles with instructions to introduce myself to Gary Felsenfeld and to learn about the unpublished, follow-up studies of polynucleotides that Gary was presenting at the Biophysical Society meeting. Professor Flory also arranged for me to lecture in his place in a special session at the 1971 Biophysical Society meeting in New Orleans. I was thrilled by the warm reception from other scientists in the program, whose studies of biopolymers were critical to my dissertation research.

Professor Flory volunteered to write letters of introduction for me to potential postdoctoral mentors when my husband was offered a position in the New York–New Jersey metropolitan area. His letter to Charles Cantor, then at Columbia University, led to a Damon Runyon fellowship and a fruitful postdoctoral year. His letter to Ulrich Strauss at the Rutgers University School of Chemistry led to my appointment as a faculty member at Douglass College, the former women's college at Rutgers. His letters of support in subsequent years were extremely valuable to my career development. I especially treasure a letter in which he discussed correspondence and scientific data relevant to our papers and encouraged me to pursue further study of polynucleotides and nucleic acids.*

> I would be interested in your comments on these points. I have no plans to reenter the polynucleotide field but correspondence, papers that appear, and the persistence vector have rekindled my interest in the subject. I am glad that you are eager to undertake calculations pertaining to

* Extract from the letter written by Professor Paul J. Flory on March 7, 1973 to Dr. Wilma K. Olson.

> formation of rings in this system. I recall your account of the effects of different bases on the ring conformation. Certainly this is an important point to pursue with respect to t-RNAs.

Personal Remembrances

My husband Gary Olson, a fellow classmate, and I were married at the end of our second year in graduate school. Professor Flory, learning of our plans, joked that he was going to have a word with Bill Johnson, chairman of the department and Gary's advisor, about the nerve of one of the Johnson students proposing without permission to one of the Flory students. He and Mrs Flory later surprised us with a memorable wedding gift—two closely fitted, deep green ceramic pieces shaped something like a beehive and presented in a box with a label from the Allied Arts Guild in nearby Menlo Park. Not having the slightest idea what this gift was and wanting to write a thank you note, we went to the Guild, a complex of artist studios and shops, looking for objects that resembled the gift. We asked the artist selling ceramics of the same color and glaze whether she had other pieces like one that we pretended to have seen there recently, and were stunned when she replied, "Oh, you must mean the cheese dish." We love to retell the story when we bring out the mystery piece to guests.

Professor Flory made up for his frequent travels by meeting with our group for afternoon tea on the days he was on campus. I enjoyed the stories that he recounted about his trips—his uneasiness in delivering a lecture at a Japanese company where the auditorium was filled to capacity by factory workers unlikely to understand English yet alone the subject matter, his descriptions of one of the first over-the-pole flights from London to California, his salmonella poisoning from a duck egg served on a flight in India. Despite his many out-of-town trips, it was not until I read an announcement in *Chemical and Engineering News*, nearly a year after entering Stanford, of the selection of Professor Flory as recipient of the 1969 Debye Award in physical chemistry that I realized how very famous he was.

I have fond memories of Flory group events, especially of two small buffet dinners in Professor and Mrs Flory's lovely home in Portola Valley. Professor Flory liked to show us the magnificent view of the mountains, bay, and evening sky from the patio by their swimming pool. I also remember him showing off Mrs Flory's collection of unusual succulents in a garden on an upper level and, when back inside the house, telling us about the unusual places where the two of them found the pine cones in their collection. Mrs Flory served delicious meals at these events and introduced me to dishes, such as English trifle, that I later prepared for others. I also recall two wonderful eggs dishes—the Chinese marbled tea eggs served at the dinner before our marriage and the creamed curried eggs that my husband so liked at the second dinner. Compared to other research group parties that we attended, the Flory events were truly the most enjoyable.

It was hard for me to hold back the tears when bidding farewell to Professor Flory upon our departure from Stanford. I saw him again only a few times and returned just twice to Stanford. My last visit coincided with the June 1985 symposium in honor of his 75th birthday. I was honored by the opportunity to present my research before him and the large Flory scientific family and I was elated by his positive comments about my work. I am so very thankful to him for setting the direction of my research career and for serving as the role model that I continually try to emulate.

Gary Patterson

Gary Patterson and Paul Flory

I was fortunate to have a very good high school chemistry teacher and a good high school physics teacher. They instilled a love of physical science and the motivation to become a scientist. They sent me to Harvey Mudd College for undergraduate school. The preparation in physics, chemistry, and mathematics was thorough. During my senior year, I read Paul Flory's classic monograph *Principles of Polymer Chemistry.* I determined to go to Stanford and work for Flory. Stanford and Paul Flory both cooperated and I started work immediately upon arrival in 1968 as an NSF Fellow.

This was a transitional period in Flory's Stanford career. Many of the brilliant experimentalists had just left, and the laboratory was full of old water baths and other obsolete apparatus. My first job was to clean the lab and design the next generation of instruments. Flory gave me my choice of thesis projects and I chose depolarized light scattering. Commercial lasers were just becoming available and Spectra-Physics developed a highly stable He–Ne laser. I built a full light scattering spectrometer and many other instruments for other students to use.

Even though I had just arrived, Flory allowed me to take his advanced polymer course. This gave me a jump-start on my research. I learned to carry out rotational isomeric state calculations of the mean-squared optical anisotropy of chain molecules such as the *n*-alkanes. Comparison with the measured quantities led to an increase in the understanding of both light scattering and the theory. Pure *n*-alkanes displayed excess anisotropy, compared with the independent molecule predictions. This effect was explained in terms of orientation correlations between the molecules. Early measurements made at Stanford were later extended and fully explained in terms of an isotropic-nematic phase transition well below the melting point of the liquid.

The 1960s were difficult times for any student. The military draft did not exempt graduate students. It was made more difficult for me since I was a conscientious objector to violence and war. Paul Flory stood up for me and

I served 2 years of alternative service. I now know that his Church of the Brethren background was pacifist. At the time I was just very grateful.

One of the greatest benefits of the Flory laboratory was the outstanding visitors to the group. During my time at Stanford I interacted with scientists like Linus Pauling and Henry Eyring. Eugene Helfand from Bell Labs had decided to explore polymer science and spent a year in the Flory lab while I was there. When it came time to look for a job, Bell Labs came looking. Both Flory and Helfand gave me a good recommendation. I spent the next 12 years in the Chemical Physics Department at AT&T Bell Labs in an office next to Helfand.

During my early career Paul Flory was a constant encouragement. The general rotational isomeric state theory of depolarized Rayleigh scattering had not been derived and the Japanese scientist, Kazuo Nagai, died prematurely. Flory encouraged me to complete the work. When I succeeded he agreed to check my work. He completed the review over a weekend. This was a major mathematical theory with hundreds of terms, mostly expressed in matrix notation. He then encouraged me to carry out the calculations. I employed the full capacity of the Bell Labs computer in this effort.

I went on to construct many dynamic light scattering instruments and to apply them to the study of polymers, both in solution and as pure materials. Paul Flory nominated me to receive the National Academy of Sciences Award for Initiatives in Research in 1981 (Figure 15.17). A picture of the ceremony is presented below. I am standing to the right of Bruno Zimm, another polymer scientist and NAS (National Academy of Science) member.

FIGURE 15.17
National Academy of Science Awardees for 1981. GDP in back row.

When the *annus terribulus* 1984 occurred and Bell Labs was decimated, Paul Flory supported me in my search for a new home at Carnegie Mellon University. While I am hardly the only person to have benefitted greatly from the continued support of Paul Flory, I am truly grateful for his constant help. His death in 1985 was a real blow.

Science is a disease, but you must catch it from someone who has it. Paul Flory was thoroughly infected. He was one of the most committed scientists I have ever met. He strove to find an understanding of observable reality. He was never content merely to play academic games. It was the truth or nothing. While even Paul Flory was not always fully correct, he never rested on his prior conclusions. He continually tried to reach the next level of understanding. He set a standard that has seldom been reached, but is a daily challenge to me.

Chris Pickles

The head of chemistry at Manchester in the 1960s was Professor Sir Geoffrey Gee. He was a polymer chemist with a particular interest in rubber physical chemistry and he was born in the same year as Dr. Flory–1910. They had a strong academic association and one aspect of this was that Dr. Flory occasionally had postdoctoral fellows from the Manchester polymer research group which, at that time, occupied the whole of the sixth floor in the department research block.

My PhD supervisor was Dr. Colin Booth who had in turn done his PhD with Professor Gee. One day in 1971 (the third and last year of my PhD) Dr. Booth showed me a letter from Dr. Flory to Professor Gee asking if he had any potential candidates for a postdoctoral fellow. I had a job offer from ICI and a steady girlfriend. A year (or more) out in California seemed exotic but also quite a risk. UK industry had already stopped the annual trawl of US-based postdocs with offers of jobs back in the UK. In the event I took my PhD viva on August 25, married Jill on August 28 and we sailed on the "Oriana" from Southampton on September 01 through the Panama Canal and docked in San Francisco on September 19. We were met by none other than Gary Patterson and his wife Sue who drove us over to Palo Alto.

To a young (I was then 24) polymer scientist Dr. Flory was about as close to a "god" as could be imagined. I had come to work on the statistical mechanics of PVC and Dr. Flory had sent me a swathe of background reading in preparation—not least his book on the subject. I had intended to spend some of the 19 days on the "Oriana" familiarizing myself with this despite the obvious distractions. Jim Mark had already looked at polyethylene so my task was to extend the application of the theory to vinyl polymers. Dr. Flory had already acquired a set of PVC [poly(vinyl chloride)] oligomers from Dr. M Kolinsky at the Institute of Macromolecular Chemistry in Prague (then the capital of Czechoslovakia).

This work was published in the *Journal of the Chemical Society*, Faraday transactions II, 1973, 69, 632–642 having been received by them on September 11, 1972. My last month at Stanford was spent preparing the submission and weekly editing meetings with Dr. Flory.

Another aspect of life at Stanford in 1971/1972 was the evolution of the use of computers in driving scientific study forward. At Manchester I had used the iconic Atlas mainframe which was the first of its kind in the world being originally developed by Tom Kilburn and his group at Manchester. The language used was called Atlas Autocode and I am probably one of a remaining small band who actually used this on the Atlas machine. The three professors Jill worked for were almost jealous of my experience. However, at Stanford the mainframe was an IBM 360/67 and so I had to learn Fortran IV to carry out my matrix computations on the statistical mechanics of PVC. There were great advantages to this—at Manchester my programs were on 7-hole punch tape and ran overnight, so it was one run per day. At Stanford we had punched cards, read them ourselves, and watched our job progress on a monitor—so several runs per day—progress indeed. Learning new software seems to have been a fact of life for over 40 years.

The Flory lab at that time was a hotbed of polymer research with at least 15 researchers operating in three connecting labs. The group was particularly social with trips to the bay islands and to Estralita's (a Mexican restaurant) among the attractions. I have subsequently kept in touch with many of the people who were there at the time. The lab was quite international in character with not only a token Brit but also Polish, Chinese, Japanese, and Indian researchers. We held weekend get-togethers and Dr. Flory hosted two great parties—one at his town house in the Palo Alto/Mountain View area and also one at his "ranch house" on the Big Sur coast range of hills (Figure 15.18).

One area of chemistry that I taught my U.S. fellow researchers during my year at Stanford was how to make English style beer. The brewing of alcoholic liquors at home for personal consumption had just become legal in California earlier in 1971 and Martin Liberman (a postdoc from the Mandelkern stable) expressed an interest in checking out the recently opened "home brew" store in San Jose. They had imported the ingredients from the United Kingdom so that was fine but they had not seen fit to offer any brewing equipment. Martin seemed pretty confident that we would be able to find a 6 gallon vat in the Chemistry Department "dead store" so off we went to have a look. Sure enough there was a vat of suitable diameter but it was double the height/depth we needed. We were prepared to cut it in half so we asked whose vat it was. Answer—none other than the double Nobel Prize winner Linus Pauling who had a lab at the end of our corridor. We went to knock on his door and he was particularly amused and happy to hand over the vat for this purpose on the understanding that he could sample the eventual

FIGURE 15.18
Flory group on an outing to Mt. Tamalpius State Park. From left: Martin Liberman, his wife and child, Chris Pickles, Larry DeBolt, Chuck Carlson and wife, Wayne Boettner; in foreground, Vincent Chang.

product. As a precaution we asked what the vat had been used for and this turned out to be the storage of dead rats in formaldehyde. At least it would be sterile! Martin continued to brew beer after my (and his) departure from the lab and 21 years later (1993) when we holidayed in California with our two young daughters we met up with him and his wife Letty and daughter Sarah and drank the very last brew he had made over 10 years earlier—he had saved two bottles for just such an eventuality.

Such was life at Stanford for me in those days—Jill got a work permit and also a post as PA to three professors in the Computer Science Department. It caused some amusement when she put her name up on the door "Jill Pickles." We saw Van Morrison perform a gig on campus and we had weekly group nights out at Magoo's Pizza in Redwood City. Nixon restarted the bombing of North Vietnam and students were avoiding the draft by either overeating or under-eating. A Stanford engineering alumnus was secretary for defence (Dave Packard). There were Nobel Prize winners aplenty including Henry Taube and Carl Djerassi (inventor of the contraceptive pill). Dr. Flory joined this group just after I left and I was lucky enough to be working in the Liverpool area back in the United Kingdom when Dr. Flory gave his Nobel address at the University of Liverpool a few years later. We were able to meet up again and reminisce together.

Dr. Flory was a most amazing person in the manner in which he carried his genius with great humility and treated all comers with great grace and respect. Although I have spent over 40 years in the chemical and related industries I am in no doubt that the year at Stanford represents the absolute pinnacle of my academic achievements and Dr. Flory was a massive part of that.

Witold Brostow

Paul J. Flory: Impressions of a Collaborator

How Did It All Start?

In describing real events, one has limitations that an author of a piece of fiction does not have. One has to stick to the facts. One also has to define a starting point, and here there are some choices. Possibly it all started when I, then a student at the University of Warsaw, tried to understand thermodynamics. I perused several textbooks. I did not like any of them, until I found a text written by Edward E. Guggenheim; in contrast to other textbooks, this one made sense. Eager to learn more, I succeeded in getting a stipend from the British Council for a stay at the University of Reading where Guggenheim was a faculty member, in fact, the chair of the Chemistry Department. Once at Reading, my mentor Maxwell L. McGlashan told me: "*Thermodynamics is incredibly badly presented, for the most part by people who do not understand it.*" He not only told me this, he also said so in an instructional article.[90]

Now comes a part pertinent to the rest of the story: Guggenheim taught a course in statistical mechanics, which I cheerfully took and learned much. Back in Warsaw, I applied my freshly acquired knowledge to a problem concerning thermodynamics of polymer chains in solution: what happens when the chain ends meet? I had derived equations for thermodynamic functions of mixing—and obtained results which Paul Flory obtained before. My derivation was different, but my final equations were identical to his; only he was not paying any attention to locations of chain ends. I sent him my publications; his reaction was: "*I did not know that my model applies to these cases also.*" He then started an action for getting me to work with him at Stanford.

From Warsaw to Stanford

Unless you are a reader of spy novels, the phrase above about "an action for getting me to work with him" sounds at least unusual. At that time the Soviet army was stationed in Poland, mostly in small towns, but in sufficient numbers to strongly influence the government of Poland. Paul Flory secured funding for my work with him, and then precisely started his action. There was an outfit in Warsaw called the Bureau of Scientific Cooperation with Abroad of the Polish Academy of Sciences. The words "Polish," "science," and "scientific" were covers; that bureau was run by Russians. Their objective was to *prevent* Polish scientists and engineers from going abroad, and above all from staying there.

I submitted the required documentation. There was an interesting part of the instructions: under no circumstances was I allowed to directly contact the U.S. Embassy in Warsaw, for instance, to ask about the current status of my visa application. Doing so would have resulted in dropping the matter by the bureau, since a scientist or engineer could not get a passport without their approval. Nobody had a passport at home; given what the authorities considered a valid reason, one was given a passport the day before departure; after coming back from abroad, one had to give the passport back within 48 h. Thus, the passport office existed, but those whose trips abroad were processed by the bureau with the long name did not deal directly with the passport office.

There was total silence after my submission of all the documentation. I had written to Professor Flory that I did not have any reply from the bureau. He wrote to them asking about the status. There was a "precise" reply that my application *"is being processed."* There were several iterations of this operation, because I still did not hear from the bureau—while Paul Flory persisted. He succeeded after 18 months. My guess now: had I asked the embassy directly, they would have said (except after 17 months) that they had not received my visa application. The bureau would have dropped my case. Since I stayed away from the U.S. Embassy as instructed, Paul Flory's insistence worked.

Arrival at Stanford

My first impressions at Stanford: everybody was so helpful! I stayed in a hotel in Palo Alto for a fortnight, and during that time I was looking for an apartment. Gary Patterson—who put the present volume together—went with me for several of my apartment searches, until I found what I was looking for.

Acquainting me with the research group, Paul Flory provided some comments—only positive—on his coworkers. I remember his remark on a student from China doing experimental work: *"he produces results at such a rate as if he had four hands."*

We were based in a building called Stauffer II. After a week or so, a friendly gentleman in a bow tie stopped me in front of the building, asking who I was, where I came from, and what I was working on. In retrospect, I think he assumed I knew who he was; this was not the case. Soon, of course, I found out: Linus Pauling. He had his office on the same upper floor of Stauffer II where I had my cubicle. Later on we talked more often, about a large variety of topics. At some point I mentioned the British philosopher and mathematician Bertrand Russell; Pauling's first reaction: *"Oh, he is a friend of mine."*

Work at Stanford

Soon enough Paul Flory suggested a research topic for me. He noted the fact that my stipend was such that I could work on anything I wished; I was not obliged to follow his suggestions. Since I came as a theorist, he suggested a topic, which for the present purposes can be defined as looking into the dark corners of the swelling of polymers in liquids. I picked his suggested topic in its entirety. It was related to the new (at the time) Flory's equation-of-state theory of mixtures including polymer solutions.[91,92] The new theory provided a large improvement over the earlier so-called Flory–Huggins theory. He also suggested to me where to find reliable experimental data, including some by his British friend Geoffrey Gee.

I was making progress in my work, when it turned out that Paul Flory had problems with his neck disk. He was wearing a stiff collar, but this did not help. Physicians in the Stanford Medical Center (the building adjacent to our Stauffer II) recommended surgery, he went for it. On the second day after his surgery, his secretary told me, "You can go and talk to him, but don't be surprised if he says something strange, he is still under the effects of various drugs." However, I found Paul Flory completely coherent. He told me among other things that for a long time physicians kept postsurgery patients on extended bed-rest, while the current thinking is that they should move around as much as they can. He had already made a short walk along the corridor earlier in the day.

When I finished my work and had written a manuscript for publication, I of course included Paul Flory as a coauthor. However, he said: because of my neck problems, I did not really help you; you did it all on your own. I argued that precise formulation of the problem means covering already half of the distance toward the solution, but he did not change his mind. Thus, the results were published with one author only.[94] Also in the swelling problem, the new Flory equation-of-state theory provided much better results than the Flory–Huggins model.

After Stanford

The Flory equation-of-state thermodynamic theory of liquids[91,92] is a general one. There are no limitations that the liquids have to be simple ones or to contain polymers. Therefore, jointly with a colleague, I have applied it to ternary liquid metal alloys.[94] As expected, the theory worked well in such systems also.

I maintained contact with Paul in various ways. There was a meeting of the Society of Plastics Engineers in San Francisco, so I went to Stanford to visit him. Because I looked for it in particular, I did locate on a wall in his office his Nobel Prize in chemistry diploma, put there as inconspicuously as possible. I was then at Drexel University in Philadelphia; at some point Paul wrote a letter supporting my tenure application.

One of his theories was the so-called switchboard model of structures of semicrystalline polymers.[95,96] Some thoughtless researcher assumed years before that at the interface of a crystalline lamella with an amorphous region, the crystalline sequences in the chains execute a perfect about-face and return into the lamella every time. This was nonsense, since the same chains went also into the amorphous regions to stay there, had loops going into the amorphous region, and also cilia (loose ends). Flory's switchboard theory pulled out the basis from all these papers. There was a bitter fight. He described to me what he called a "scurrilous" behavior of a drunk supporter of the about-face model at an international meeting in Russia. Elsewhere I supported him, pointing out among other things that the regular about-face model violated the laws of probability.

1985

In June 1985 there was a 3-day celebration of Paul Flory's 75th birthday at Stanford. Four colleagues put together three volumes of his *selected* papers.[97] Many of us participated in that birthday event.

It so happened that later in the same year I followed his trail. On Friday, August 9, he gave a lecture at the Johannes Gutenberg University in Mainz, then the leading center of polymer research in Europe. I gave my lecture there the following Monday; both announcements were on the same blackboard, separated by a horizontal line. On Monday, August 19, Paul gave the opening lecture at the 30th Macromolecular Symposium of the International Union of Pure and Applied Chemistry in The Hague. The following day he came to my talk. There was a reception for the participants in the Knights Hall of the moat-surrounded royal castle. He and I talked at the conference and at the reception about a number of things: free volume, Dutch architecture, and crank mail he was receiving in large quantities. We also talked about giant redwoods at Muir Woods and about some pines that grow along the California–Nevada border. In both cases, a tree falls but a new one grows from the same roots. Is this a new tree, or a different one? We were not able to decide.

On August 21 my wife Anna, my children Gabriel and Diana and I took a flight from Amsterdam Schiphol to New York. Paul Flory was on board, and we talked about ordinary things: about the plane delay, why and for how long. We had no idea we were seeing him for the last time.

After September 8, the news of Paul passing away reached us. I told this to Anna, Gabriel and Diana. About the same time news of passing away of a cousin of mine reached us too. Diana said: "See, your cousin just died and your professor too." Somehow, in her 5-year old mind, Paul was regarded as a member of our family. Since, apart from everything else, Paul was also a member of the Editorial Board of Materials Chemistry and Physics, I wrote a remembrance of him for that journal.[98]

In 1977 Andrew M. Sessler of Lawrence Berkeley Laboratory created an international organization called Scientists for Sakharov, Orlov, Sharansky

(SOS) to help scientists around the world (largely in the Soviet Union) persecuted for political reasons. Paul Flory was an active member of SOS; I participated, among others, in a meeting of this organization in New York City after September 8, 1985. Sakharov's wife was present, freshly released from the Soviet Union, while her husband remained expelled from Moscow and lived under strict control in Gorky (now Nizhny Novgorod). Speaking for myself and on behalf of fellow Flory students, I said: "We shall continue Paul Flory's activities in SOS."

A Final Thought

Some people divide research into "basic" and "applied." I think this is wrong. Paul Flory has taught us that, to solve "applied" problems, one needs to do "basic" research first, to understand the phenomena, behavior and/or processes involved.

Do Y. Yoon

My Reminiscences of Paul J. Flory

It was my real good fortune to join Paul J. Flory's group as a postdoc in July, 1973 after finishing my PhD thesis under Richard S. (Dick) Stein in the Polymer Science and Engineering Program of University of Massachusetts at Amherst (UMass). Paul offered me a postdoc position to work on spatial configurations of macromolecular chains, based on his conversation with Dick during his lecture visit to UMass in October, 1972, although I had no background in this topic since my bachelor degree was in chemical engineering from Seoul National University, Korea and my PhD thesis was on the light scattering and deformation characteristics of semicrystalline polymers. Therefore, Paul's offer was a big, pleasant surprise but at the same time a huge scare for me, but Dick convinced me to accept the offer since Paul was the best theoretician in polymer science at the time and I could learn lots of new things from him. I am still deeply indebted to Dick for his foresight and encouragement for me to take a new path in my career.

My first project in Paul's group was to calculate the higher tensorial moments of chain persistent vectors in order to test the applicability of Paul's new theory of describing the non-Gaussian nature of the chain vector distribution of real finite-length chains. Although the mathematical formulae were rather straightforward, the actual computations of very high moments, averaged over all possible rotational isomeric states, required an enormous amount of computer memory and computation times for matrix operations, which were beyond the limits of the Stanford computing center. Therefore, Paul was very delighted and impressed when I came up with a practical computational solution by programing the algorithm instead of the mathematical formula, and included the newly obtained results as part of his Priestly Medal lecture in the 1974 spring ACS meeting. This early success I had in solving computational problems prompted Paul to suggest other challenging problems that he was concerned with at that time, including the conformational characteristics of polystyrene, poly(methyl acrylates) (PMMA), and so on, with all resulting in satisfactory outcomes.

One memorable case was the puzzle in interpreting the neutron scattering results of amorphous PMMA at intermediate scattering vectors. By early 1974 all the experimental results of the small-angle neutron scattering (SANS) from the mixtures of normal hydrido-polymers and their perdeuterated counterparts, just carried out in Europe for the first time, convincingly demonstrated the validity of the unperturbed Gaussian random-coil model, that Paul first predicted in 1949, upon comparing the radii of gyration (Rg) measured in bulk amorphous melt/glass state with those measured by light scattering in dilute theta solutions. However, the results of intermediate-angle neutron scattering (IANS) pattern of PMMA, that probes the local conformational structure below Rg, showed a strong deviation from the well-known Debye equation predicted for Gaussian random coils. And this deviation was highlighted to support the locally "ordered" chain model of bulk amorphous polymers advocated by others. So, in the spring of 1974 Paul strongly suggested that I look into this problem since I was familiar with scattering phenomena due to my PhD thesis work. In early summer I showed Paul a plot of characteristic ratio of syndiotactic PMMA (s-PMMA) chains versus the chain length, which showed a maximum before reaching a plateau value at high chain lengths (see Figure 15.19), in contrast to the

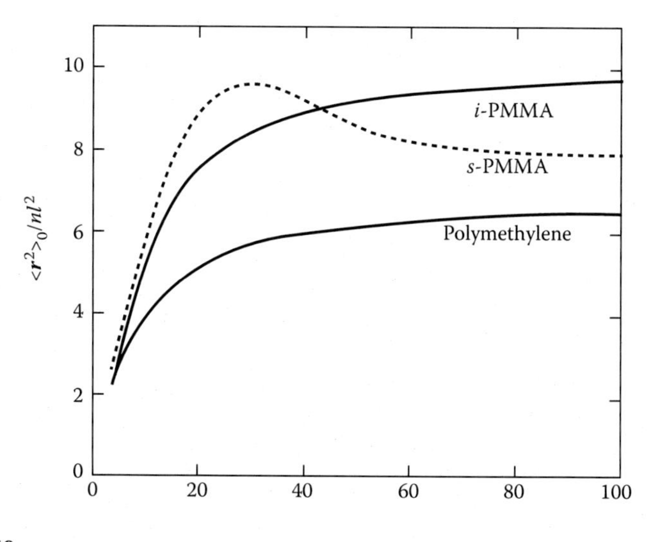

FIGURE 15.19
Plot of characteristic ratio versus skeletal carbon length n.

general behavior of gradually increasing characteristic ratio values to reach a plateau, observed for practically all other polymers. I mentioned that this could explain the anomalous IANS pattern observed for PMMA. Paul got really excited to see this result, immediately agreed with my proposition, and urged me to complete the full calculation of IANS patterns for PMMA chains of varying stereoregularity in comparison with other polymers such as polyethylene, polystyrene, and so on. Indeed, the calculated IANS patterns of random-coil syndiotiactic PMMA and atactic PMMA chains showed the experimentally observed deviation from the Debye equation.

Paul was really pleased with this work on IANS patterns of real polymer chains since it dispelled any remaining doubt on the validity of the unperturbed Gaussian coil-model of flexible polymers in bulk amorphous state over the entire chain-length scale. He announced the preliminary results of this work in the special polymer symposium held on the Stanford campus in July, 1974. And when he was awarded the 1974 Nobel Prize in Chemistry and was preparing for the Nobel award lecture, he asked me to prepare a special figure so that he could highlight this result as part of his Nobel lecture and proceedings paper in December, 1974 (Figure 15.20).

In the fall of 1974, I discussed with Paul about my future career plan. He strongly advised me to look for an industrial or government research position in the United States since I would encounter great difficulty in an American academic institution due to my lack of English skills in proposal writing and teaching, and offered his help in contacting potential employers. I am still very grateful for the efforts he made on my behalf in contacting a number of prominent industrial and government research laboratories. One of those places he personally contacted was the IBM Research Laboratory in San Jose, about 40 miles south of Stanford University, and his

FIGURE 15.20
With Paul at the 1974 Chemistry Nobel Prize reception on Stanford campus (October, 1974).

strong recommendation (as a new Nobel laureate and also a member of IBM Science Advisory Committee) was a big help for me to receive a very attractive offer in March, 1975 as a research staff member in the Polymer Science and Technology Department. Paul was very pleased to hear the good news and mentioned that he would be very interested in continuing the research collaboration with me since IBM Research encouraged such collaboration.

One area of potential collaboration we discussed before I joined IBM in August, 1975 was the modeling of neutron scattering patterns from semicrystalline polymers since the experimental projects were underway in Europe, and we agreed that the type of conformation-based modeling of IANS patterns we carried out for amorphous polymers could provide very valuable information on the molecular morphology of semicrystalline polymers. At that time the issue of a regular folding model with adjacent reentry (supported by many experts in crystalline polymers) versus the random switch-board model with nonadjacent reentry (supported by Paul and Leo Mandelkern) was still very much in big controversy since the 1960s and neutron experiments combined with molecular modeling had the potential to resolve this controversy. Paul proceeded to get in contact with the key experimental groups (George Wignall and his coworkers) to encourage them to measure the IANS data and to send us the available data as quickly as possible.

We received the first set of neutron scattering data on semicrystalline polyethylene in the fall of 1975, which surprisingly showed practically the same scattering pattern as that of polyethylene melt chains, not only in the values of radius of gyration but also in the IANS pattern. Paul and I discussed the modeling strategy by recognizing the lamellar semicrystalline morphology of crystallized polyethylene samples, with the molecular variable being the rotational isomeric state model of polyethylene in noncrystaliine region and the stochastic probability of adjacent reentry at the crystal–amorphous interface. Upon comparing with experimental results for rapidly crystallized polyethylene samples, we came to a definite conclusion that the probability of adjacent reentry is less than 30%. This work, completed with the computing resources in IBM Research and first published in early 1977, generated a lot of debate/controversy related to the modeling methodology that resulted in anomalous density in the crystal-amorphous interfacial region; the effect of this deficiency on the IANS pattern is very minor as later calculations with the correct density model showed.

As more experimental results became available for other semicrystalline polymers, including slowly crystallized isotactic polypropylene, a very surprising conclusion was emerging that the overall radius of gyration of a polymer chain in polymer melt remains nearly the same upon crystallization from the melt. Paul mentioned that this was a most significant scientific result and began to raise the fundamental question on the molecular morphology in semicrystalline polymers. He asked me to look into the longest relaxation time (disentanglement time) of a single polymer coil in highly entangled polymer melts versus the time its radius is traversed by the growing crystal front. When the time scales of the disentanglement time turned out to be much longer, by a factor of ca. 100, than the time interval of the crystal growth passing through the spatial dimension of a single Gaussian coil in highly entangled polymer melts, Paul became convinced of a new model of molecular morphology of melt-crystallized semicrystalline polymers. That is, the rapidity of crystal growth along the growth front planes, once nucleated, allows only the straightening of one crystalline stem length of chain sequences locally within a long random-coil chain, leaving the overall radius of gyration and the resulting highly interpenetrated (entangled) molecular morphology nearly unchanged upon crystallization from the melt. Paul drafted the manuscript for submission as a *Nature* article on his flight to Seoul for a series of public lectures in Korea in the fall of 1977. The paper was first presented by me as an invited talk in the 1978 spring APS meeting, in a special session organized by Eugene Helfand, and was followed by the hottest emotional debates ever seen, and its publication in *Nature* in 1978 also generated heated discussions. As a result, Paul and I were invited to participate in the Faraday Discussions of the Chemical Society on amorphous and crystalline polymers held in Cambridge in the fall of 1979, which was another highly charged discussion meeting, well documented in the published proceedings of Faraday Discussions. This emotionally charged meeting experience of the chain-folding controversy was very tiring and Paul felt deeply disappointed with the rather overly emotional responses that our work generated, as expressed in his own summary remarks in the proceedings. Despite this disheartening experience, Paul was motivated to rigorously solve the packing density problem at the crystal-amorphous interphase of lamellar semicrystalline polymers by employing a new lattice model in collaboration with Ken Dill and me.

Paul formally retired from Stanford University in 1977 in accordance with the mandatory retirement policy of the university at that time. His continuing interaction with senior IBM management as a member of the IBM Science Advisory Committee and the desire of the IBM Research management to expand/upgrade the polymer research programs prompted IBM Research to hire Paul as a senior consultant to the IBM San Jose Research Laboratory spending 2–3 days a week from 1977 until his death in 1985 (Figure 15.21). During these 8 years, IBM provided Paul with 2–3 postdocs/visiting scientists each year who worked in San Jose Research Laboratory under his supervision, belonging officially to my polymer physics group. This provided an ample opportunity for me and other IBM colleagues to interact with Paul and his associates for a long time. As a senior consultant to IBM Research and a member of the IBM Science Advisory Committee, Paul strongly promoted and supported polymer related advanced technology projects in IBM development laboratories in United States and polymer research projects in the San Jose Laboratory, especially working with Jim Economy who built up a very successful Polymer Science and Technology Department.

I started a 1-year research sabbatical in the Max Planck Institute for Polymer Research (MPIP) in Mainz, Germany from July, 1985, and Paul visited MPIP for 2 days on his way to the IUPAC meeting in The Hague, Netherlands in August, 1985. Prior to visiting MPIP, Paul spent 5 days in the Austrian Alps participating in an IBM Europe Institute symposium on polymer science. There, he enjoyed hiking together with younger scientists and very actively participated in all the symposium activities. His schedule in MPIP was very busy but he seemed well relaxed and fit on the day he left for The Hague. So, it was a tremendous shock for me and everybody at MPIP when about 3 weeks later I received a phone call from Tom Russell in San Jose, informing me that Professor Paul Flory had a heart attack and passed away at his cabin in Big Sur. In a week, I returned to the Bay Area and visited Mrs Emily Flory and learned that Paul had gone to the Big Sur cabin on the weekend to work on his invited talk to be delivered to the ACS fall 1985 meeting the following week, since he always slept better at Big Sur.

So, Paul was working very hard, perhaps too hard, till the end. Looking back, Paul was the hardest working scientist I ever met. Although Paul and Emily spent a considerable amount of time together on the human rights issues of Soviet scientists after he received the Nobel Prize in 1974, he published about 120 papers between 1975 and 1985, as compared with about 240 papers published up to 1974. His intuitive, scientific insight was truly extraordinary, but his dedication to the scientific truth, discipline, hard work, and time-consuming devotion to the rigor and to the details was second to none.

In the spring of 1985 Paul seriously considered updating his famous book *Principles of Polymer Chemistry* published in 1953, and he mentioned that he would definitely add a new chapter on "Molecular Morphology of Bulk Polymers," which I wholeheartedly agreed to. Then, in about 1 month he told me that he decided against the book project because it would take almost two full years to do a good job, which would be too much time taken away from his ongoing research work. Until the end he still had many new exciting ideas to pursue!!

One of the new projects he was really excited about was described in a draft manuscript Emily found in Paul's briefcase, entitled "Conformational Rearrangement in the Nematic Phase of a Polymer Comprising Rigid and Flexible Sequences in Alternating Succession. Orientation-Dependent Interactions Included." Emily passed Leo Mandelkern a manila folder that contained the draft manuscript with a title and a brief sketch of the theoretical model and the derived equations together with a several notebook pages of calculated data tables, when Leo came out to Stanford to sort out Paul's office files after his death. Emily told Leo that Paul regarded this work to be very significant and expressed her wish to have it published in the original form only. However, the draft was too rough with no description of calculated results, and so on. Therefore, Leo sent me a photocopy of the whole package and asked me whether I could figure out the content and possibly make it into an acceptable manuscript. After I returned from my sabbatical,

FIGURE 15.21
With Paul at IBM San Jose Research Laboratory (May, 1985).

I studied Paul's final manuscript, understood what he had intended to calculate, and completed the remaining calculations to my satisfaction. After some discussions among Emily, Leo, and me the work was published in the Proceedings of the Materials Research Society Symposium on "Materials Science and Engineering of Rigid Rod Polymers," organized by Wade Adams and dedicated to Paul in 1989: *Mater. Res. Soc. Symp. Proc.*, 134, 3&11 (1989). I still wonder what Paul would have thought of the final results and the discussions/conclusions I added.

Paul Flory has been my mentor, advisor, and role model ever since I first met him in 1973. While at IBM Research Laboratory and later at Seoul National University, I tried to pass along Flory's way of doing science to all the postdocs and graduate students in my research group. Even now at Stanford University, my undergraduate and graduate courses on polymers would not be complete without enough doses of Flory's theories and discoveries for the benefit of mankind.

Burak Erman

Ten Years with Flory

I joined Professor Flory's lab in the summer of 1976 for 2 years as a postdoc. My educational background was engineering, with particular interest in continuum mechanics and elasticity. When I arrived at the lab, Flory was developing the model of rubber elasticity with constrained junctions. My background was suitable for this project and within a month after my arrival I started experiments on the small deformations of PDMS networks both under tension and compression. The aim was to see whether the elastic modulus passed smoothly from the tension to the compression region. There were suggestions at that time that the modulus should diverge at small deformations, which was not predicted by the molecular theories of rubber elasticity. Following the experimental work, Flory asked me to work on the theoretical model on the effects of constraints on junctions. We published several papers in this field until 1985, the year Flory died. Around 1975, or perhaps several years before, Flory had already realized that the elastic moduli of polymer networks changed from larger values at small deformations to smaller values at large deformations. His idea was that junctions are severely constrained by their environments and stretching them or swelling them with a solvent allows more space to them to fluctuate in, thus exhibiting a softening effect which causes the moduli to decrease upon stretching or swelling. This view of molecular structure now forms the basis of protein physics, but it was not appreciated much when Flory was first suggesting it in the seventies. The importance of fluctuations in rubber elasticity was actually proposed a year or two before by two Italian scientists, Giorgio Ronca and Guiseppe Allegra in a brilliant paper which was the major reference in Flory's famous Constrained Junction Model published in 1977. Soon after this paper, we published another paper on the geometrical nature of constraints in rubber elasticity. After the Ronca-Allegra paper was published, Giorgio Ronca spent 2 years at Stanford as postdoc to Flory. However, they worked not on rubber elasticity but on the statistical mechanics of liquid crystals. At the time when Flory was developing the constrained junction model of rubber elasticity, there was another group of scientists suggesting that the observed excess elastic moduli were due to trapped entanglements. The two opposing views formed the basis of a strong controversy which was never resolved but simply died down over the years. Flory was very strict in his view, always saying that whatever was seen in excess of the equilibrium statistical mechanics was due to incomplete relaxation of the samples. At the end of my postdoc I was returning to my university in Istanbul and Flory asked me if I could come back to Stanford the following year during the summer. I was very happy with this offer, and my trips between Istanbul and Stanford continued until Flory's death in 1985. In total, I made one sabbatical year and six summer visits to Flory's lab. Throughout those years, I realized over and over how great a scientist Flory was. During one of my visits, I was interested in calculating the higher moments of polymer conformations, which he asked Enrique Saiz and me to work on. Enrique and I spent days and nights in calculating the moments using the main Stanford computer. We then proudly went to Flory's office to present our results. He looked at them for a few seconds, made a few back of the envelope-type calculations, and said he would expect one of the moments to be about 5% smaller. With Enrique, we went back to the Stanford computer again, spent several more days, and realized that we had a small mistake in the program, which changed the value to a number very close to what Flory expected.

Flory's research group was always small, five or six postdocs and a few visitors. In 1976, Wayne Boettner, Flory's last student, was about to receive his PhD. Gary Patterson had already finished his PhD with Flory and left for Bell Labs. His results on scattering were of continued interest to Flory, which I applied to scattering from the bulk state. Ken Dill arrived from Bruno Zimm's lab for a postdoc and started working on the statistical mechanics of bilayers. Enrique Saiz was the postdoc from Madrid who was already at the lab when I arrived in 1976. He visited for another summer later on. Wayne Mattice from Louisiana and Don Napper from Australia were visiting scholars for a year. Do Yoon had already finished his postdoc at Stanford and took a permanent job at IBM Research at San Jose. Flory spent 2 days per week at IBM Research and 3 at Stanford. Ronca was working on liquid crystal theory and moved to IBM. Later on, I got interested in aromatic polyesters, and started to spend 2 days per week at IBM and 3 at Stanford.

Flory and Mrs Flory, Emily, often invited the group to their house in Portola Valley, where we swam in their pool and Flory barbequed hamburgers, served with white or red California wine. The following Figures 15.22–15.26 reflect the many happy times we spent at their Portola Valley home.

We were all very curious about the Flory house at Big Sur. Nobody had been there. We knew that it was in the wilderness, at the top of a mountain named the Flory Mountain, which truly was written exactly like that on an official map I saw. One weekend, Flory and Emily invited the group to the Big Sur house. We were all very excited. We could drive only halfway up

FIGURE 15.22
(Left panel) discussing science after a swim at the Flory house in Portola Valley, Burak Erman (center), Dacheng Wu (right). (Right panel: left to right) Yuri, Flory, Bob, Burak, Wu, and Peter. Flory made a somersault from the springboard into the pool and Emily was not very happy because Flory had a neck operation and was supposed to refrain from stressing his neck.

FIGURE 15.23
The Flory group at Flory house in Portola Valley. (Standing from left to right) Dacheng Wu, Bob Orwoll, Michele Vacatello, Flory, Milenko Plavsic, and Yuri Yarim Agaev. (Seated, left) Peter Irvine, (right) Burak Erman.

the mountain and Flory came and picked us up with his four wheel drive from where we could not drive further. It was like a nightmare going up the mountain, not because of the way Flory drove, but because the road was so tricky that we had to take several attempts at the turns, maneuvering back and forth each time, with the back of the car hanging almost out over the

FIGURE 15.24
Emily cutting the cake by the pool and Wu watching, wine glass in his hand. Portola Valley is seen in the background. The well-trimmed hedge made the pool a private place where no one from the valley could see the swimmers.

FIGURE 15.25
Peter Irvine in the sunny Portola Valley evening, discussing with Flory ways of melting sexaphenyl liquid crystal molecules and observing phase transitions.

cliff. The house was beautiful, overlooking the Pacific Ocean. At the door as we entered the house, was written: "Solitude Without Loneliness." We were a group of seven, Bob Orwoll, a former student of Flory, visiting for the summer, and his wife, Dacheng Wu from Beijing whom Flory invited as a postdoc at the end of the Mao regime, Milenko Plavsic, postdoc from Belgrade, myself, and my wife Gulden and my son Batu.

FIGURE 15.26
Dinner after the hike.

FIGURE 15.27
Flory explaining to us how the pillars of Milet stayed intact for more than 2000 years.

The women stayed with Emily, and Flory took the rest for a hike over the Sierra Mountains. We first climbed down a very steep hill behind the house, then up a steep hill, down and up a couple of times, being very cautious not to get caught by poison ivy. Flory was leading the group, and a moment came when Flory stopped and said, "I don't have any idea where we are." It took us more than 3 h to reach the top of a hill. We could see the house about 5 miles and several hilltops away. Wu said he could run to the house over the hilltops and have Emily come with the jeep. Flory was skeptical that Wu could do that. Wu said he used to commute to the university back home, 1 hour each way, all year round. He ran and told Emily where we were and she rescued us.

The times I spent at IBM San Jose were the intellectually most stimulating years of my life. Flory was always available for discussion. One summer Ueli Suter visited from ETH Zurich. He and Flory were working on the exciting problem of conformations of poly(isobutylene). Akihiro Abe spent one summer at IBM with Flory, working on liquid crystal forming nematics. They produced a series of the most important papers at the end of that visit. At the same time at Stanford, Bob Matheson, Peter Irvine, and Matthias Baulloff were postdocs working on nematic-like systems, all doing theory as well as experiments. The postdoc Yuri Yarim Agaev, a Russian dissident, whom Flory helped leave Russia, was also at Stanford. Flory, although essentially a theoretician, had great insight into experimental issues. One day he asked me to go to Stanford because Peter Irvine was trying to dissolve sexaphenyl in boiling dimethylsulphoxide (DMSO), and he felt that there should be two people in case something went wrong. And sure enough, as soon as I arrived at the lab at Stanford I saw Peter in front of the hood, the hood in flames because the DMSO had boiled over and caught fire from an electrical spark. Peter was trying to keep the setup steady and unable to leave to call for help. There were special powder fire extinguishers along the corridor that came to our rescue. Other than that incident, I never saw Flory interfere with experimental processes in the lab. One day he came to the lab to discuss a problem, leaned against the central table and felt the warmth of an oven under the table, which he said must have been on for several years after one of his earlier PhD students using that oven had left.

I loved to discuss things outside of science with Flory. He was very good at imitating people, but only when people close to him were around. Once deGennes said that Flory's excluded volume exponent was correct but it was because of the cancellation of two errors. During one of our trips from IBM Research at San Jose to Stanford, I told this to Flory. He was silent for a few minutes and then said "deGennes is so charming that even I cannot get angry with his nonsensical remarks."

In the spring of 1983, Flory and Emily visited us in Istanbul for 10 days. Four of us, Flory, myself, Emily, and Gulden took a tour of the ancient Greek sites along the Aegean coast. Flory was very knowledgeable about the places we visited. Our first stop was Troy where we saw the gate through which Schliemann took the gold. In our second stop, Pergamon, Flory told us the story of the engineer who brought water to the city, and on hearing the first drops of water trickling into the well he jumped up in happiness and lost his balance and fell down the cliff and died. Flory was very excited about seeing Milet, the city where Thales lived. It was equally exciting for me to see Flory in the land where Thales lived 2500 years ago. I always thought of Thales and Flory being equal contributors to civilization, and there we were standing in the barren soils of Milet with one of them (Figure 15.27). We then visited Priene, and Flory said it was as if something happened 2000 years ago and people of the city locked their doors and left in a hurry. On our way to Ephesus, our next stop, I read about Joel Hildebrand's passing away and told this to Flory. Flory was deeply saddened. After several minutes of silence, he said "I wouldn't like to be in his place during his last years."

In 1985, I went to Stanford to spend a sabbatical with Flory. We were working on the problem of critical phenomena and phase transitions in gels. We often had lunch together at IBM, with Do Yoon and Ueli Suter. Flory used to go to Big Sur more often. It was the first time when I saw him being emotional about watching the migrating whales from his house at Big Sur. I even remember him describing them as "moving gracefully to eternity." We finished our work on phase transitions in gels and he was going to present the work in the Chicago ACS Meeting. On the Friday before the meeting, he said he wanted to concentrate on the talk and went to Big Sur. He was supposed to come back on Sunday and then leave for the meeting. He never came back.

He had a massive heart attack at Big Sur. He felt that there was something wrong with his heart because later it was found out that he stopped by the local pharmacist at Big Sur and bought cough pills. I always thought that it was typical of Flory. He experimented and watched for the results and then formed a model, which one sees over and over in following his works. But this last time the experimental result was negative. Akihiro Abe called me on Monday morning to see whether what he heard on the radio in Tokyo was true. He was the first one to hear of Flory's Nobel Prize and called to congratulate him, and he was the first one to hear of his death.

Flory was a great mentor even when he did not say anything, because we all knew that he was after the "universal," and working with him, for him, gave us that feeling and that was the greatest intellectual experience.

Ken A. Dill

Reminiscences about Paul Flory

I first met Paul Flory around 1977. I was a graduate student with Bruno Zimm at UCSD at that time, and Zimm suggested that I consider postdoctoral work with Flory. Flory had been invited to give two special lectures at UCSD, and Zimm had kindly arranged a time that I could meet with Flory during his visit. My first impression was of Flory's lectures. I found his talks to be mumbling, uninspiring and quite dry. Flory's style was to write out his talks in advance, word-for-word by hand. Then, for the lecture, he would read his notes verbatim, mostly staring at his shoes. Nevertheless, I was delighted when, during our chat afterwards, he indicated a willingness to take me on, mostly I suppose, based on Zimm's word.

Upon my arrival in his lab in 1978 as a postdoc, it did not take long to see that Flory was larger than life. I asked him what he wanted me to call him. He said: "Call me Paul." Given the California culture, that made sense. But I soon discovered that everyone else called him "Professor Flory." He was viewed with great reverence and held in the highest respect by all around him. Much of the time, he wore a coat and tie. He had a commanding way about him and a clear sense of purpose. For example, I had arrived—post-doctoral fellowship in hand—with the agreed purpose of working on the physical modeling of the chain packing inside micelles and bilayers. But much to my surprise, about a week after I arrived, Flory said, in effect: welcome to the lab, forget about micelles, now let me tell you about a nice little liquid-crystals project I have in mind for you! (In the end, he acceded. And, later, I was delighted that, when he was called upon to deliver the inaugural Paul J Flory lecture in 1984 in the Stanford Chemistry Department, he chose our work on membranes and micelles as his topic.)

His writing style was old-school and formalistic. It took me several years after my postdoctoral stint to unlearn to write papers in his voice, with long sentences having complicated conjunctive clauses, "heretofore's," "inasmuch as's," and so on.

Paul had strong opinions and an unyielding personality. He did not suffer fools gladly. From my postdoc office, I could see his office door, and once in a while I would see a visitor almost shoved out of his office, if Paul became impatient with the visitor. The 1979 Faraday Discussion of the Chemical Society was a battleground meeting between two camps. On the one hand, Sir Charles Frank's camp believed that semicrystalline polymer interphases (the intervening chain regions between crystalline regions of polymers) were organized like the regular weave of a fabric. Flory believed it was more disordered and statistical. That meeting was quite vitriolic, with name-calling and people marching out of the room in disgust. Some of that flavor is reflected in the published proceedings. It has been said that the contentiousness around the field of semicrystalline interphases was more a reflection of the personalities of Flory and Frank than of the scientific issues, *per se*. One further amusing story. Paul once gave a manuscript to his secretary, Doris, a very sweet North Carolina woman, that had a small grammatical mistake in it—something like putting a comma outside of quotation marks, rather than inside. Doris, ever the conscientious secretary, fixed the mistake after consulting her secretary's manual. But Flory insisted that he was right, and that the manual was wrong, and that it should be typed exactly the way he specified. So Doris sent the manuscript to the journal Flory's way. Sure enough, when the journal sent back the proofs, it had been corrected back to Doris' way.

I want to comment on two ways that Flory had a huge influence on my life, that were also prescient predictions of our modern understanding of protein molecules. Paul was very interested in proteins, but his thinking was decades ahead of his time. Paul is well known for his role in proving that polymers are long-chain states of matter, and his role in the revolution away from the view that molecules were only pure and understandable if they had single crystallizable structures and toward today's view that properties of molecules and materials cannot be understood without statistical physics. His classic 1953 book, *Principles of Polymer Chemistry*—particularly the first chapter—remains a masterful tale of the history of this transition. Now, rolling the tape forward, modern protein science too has discovered the power of statistical physics, entailing a similar transition. Early protein science had a heavy emphasis on the importance of native states, atomic details and well-defined particular structures. The more current perspective is that proteins, like synthetic polymers, are heavily governed by their statistical physics—their large entropies, their often-extensive degrees of intrinsic disorder and their transitions, such as the folding process, that entail considerable microscopic heterogeneity.

Second, Paul was a master at seeing the simplest essence of a problem and developing deeply incisive analytical theories. For me, one of his deepest and most critical insights was excluded-volume, which, for my purposes here can be expressed most concisely as (z/e)^N (PJ Flory, Thermodynamics of high polymer systems, *J. Chem. Phys.*, 9, 660, 1941; see Equations 15.1 and 15.3). If z is the number of rotational isomers of a bond, and if there are N bonds, then z^N estimates the size of the space of unhindered conformations, while (1/e)^N, introduced by Flory, represents the huge reduction of conformational space of chains that are constrained by surrounding chains in dense-chain media. This factor of e^N later became critical for our discovery that proteins fold via funnel-shaped energy landscapes (KA Dill, Theory for the folding and stability of globular proteins, *Biochemistry*, 24, 1501, 1985). As a protein is folding, it becomes increasingly dense, which reduces its conformational space by such a factor. Interestingly, Flory had previously guessed that some simple excluded-volume argument like this might apply to the folding of proteins (see PJ Flory, *Statistical Mechanics of Chain Molecules, Wiley Interscience*, 1969, pp. 302–303).

I have always deeply appreciated Paul Flory's role as a wonderful mentor to me. After I left his lab in 1981, I pursued statistical mechanical theory of protein folding. Until his death in 1985, I returned regularly to Palo Alto and talked to him about it. My theory was showing that folding led to a huge reduction of conformational space, but not to a single native structure. But Paul kept pushing because he believed we should be able to go further, to understand why native proteins had single structures. His pushing led us to develop the HP lattice model, which allowed us to test errors in mean-field modeling through the use of exact enumerations (KF Lau and KA Dill, *Macromolecules* 22, 3986, 1989). Indeed, Flory was right: when we went beyond mean-field approximations in lattice models, we found that binary-sequence foldamers with excluded volume would indeed often fold to single native structures.

Paul Flory was deeply insightful. He had incredible talent for extracting the simplest essence of a problem. He could develop exactly the right level of model that could be analytically tractable on the one hand and make meaningful predictions about nature and experiments on the other. He had extraordinary facility with probabilities and combinatorics. He had great breadth that spanned theory and experiments. And, to the people he valued, Paul was warm, caring, and supportive. Moreover, perhaps because of his upbringing as a clergyman's son, Paul greatly valued modesty in himself and others. Remarkably, the author index in the back of Flory's 1953 book, *Principles of Polymer Chemistry*, scrupulously and extensively cites all the authors whose work is referenced in his book, with one exception—there are no references to Flory himself! And, he once told me that no CV should ever be longer than two pages. Indeed, his own CV was a masterpiece of modesty: the second of its two pages was a list of awards, and alphabetically listed under "N" were: Nobel Prize, National Academy of Sciences and National Medal of Science.

16

Honors and Awards

Paul John Flory received many honors and awards. As noted in the chapter on Ohio State, he received his first medal, the Joseph Sullivant Medal, in 1945. This was a career award for international distinction, but Flory received it after only 11 years. Ohio State also awarded Flory a DSc in 1970.

One of the local ACS Awards that is especially distinguished is the Leo Hendrik Baekeland Award of the New Jersey Section (Figure 16.1). Paul Flory received this honor in 1947, the second person to do so. Many Baekeland awardees have gone on to receive the Nobel Prize. The award is restricted to scientists under 40 years of age.

One of the highlights of his year at the University of Manchester was the Colwyn Medal of the Institution of the Rubber Industry in Great Britain in 1954. The award was named for the Right Honorable Lord Colwyn PC and the first awardee was G. Stafford Whitby in 1928. Geoffrey Gee received the Colwyn Medal in 1952! The Victoria University of Manchester lauded Flory with a DSc in 1969.

Another classic award was the Nichols Medal for 1962 (Figure 16.2), sponsored by the New York Section of the ACS. It is given for original chemical research and former recipients include many later Nobel Prize winners such as Irving Langmuir, Linus Pauling, Glen Seaborg, and Peter Debye. Former winners for research in polymers include Leo Baekeland, Carl Marvel, and Herman Mark.

The High Polymer Physics Division of the American Physical Society awarded its High Polymer Prize to Paul Flory in 1962.

Flory was appreciated in both the industrial and the academic worlds. The Society of Plastics Engineers gave Flory the International Award in Plastics Science and Engineering in 1967 as part of the 25th anniversary meeting of the society.

Paul Flory's contributions to rubber science were recognized by the Charles Goodyear Medal of the Rubber Division of the American Chemical Society in 1968. After his death he was elected to the Rubber Science Hall of Fame in Akron, Ohio. The list of Goodyear medalists and Hall of Fame honorees is very similar (Figure 16.3).

Paul Flory was also recognized as one of the great physical chemists of the twentieth century by the awarding of the Peter Debye Award in Physical Chemistry of the American Chemical Society. Paul self-identified as a physical chemist and was always proud of the area of his doctoral degree.

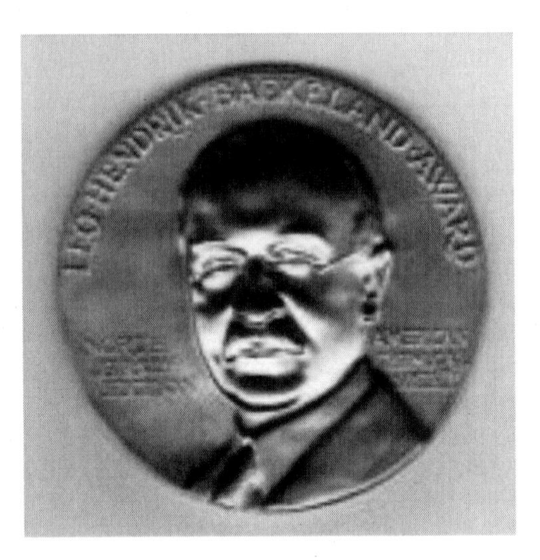

FIGURE 16.1
Baekeland Medal.

One of the founders of the American Chemical Society and its president from 1881 to 1889, Charles Frederick Chandler, was memorialized by his friends with a medal given to him in 1910 and to many famous chemists in later years. Flory received this award in 1970. It reflects the perceived breadth of Flory's interests in chemistry and the wide influence his work had on the whole field. Leo Baekeland was the second such recipient in 1914.

FIGURE 16.2
Nichols Medal.

FIGURE 16.3
Charles Goodyear Medal.

One of the oldest medals awarded for international science in America is the Elliott Cresson Medal of the Franklin Institute in Philadelphia (Figure 16.4). Paul Flory received this medal in 1971. He joins polymer scientists Jacques Edwin Brandenberger, Joseph C. Patrick, Waldo Semon, and Herman Francis Mark in receiving this medal (Figure 16.4).

Another great physical chemist that influenced Paul Flory's career was John Gamble Kirkwood (1907–1959). In 1962 the New Haven Section of the

FIGURE 16.4
Elliott Cresson Medal.

FIGURE 16.5
J. Willard Gibbs Medal.

American Chemical Society and Yale University established a prize in his name. Paul Flory won this award in 1971. More than half of the awardees also received the Nobel Prize soon afterwards. Henry Taube won it in 1966 and Bruno Zimm was the awardee for 1982.

The Willard Gibbs Award was established in 1910 by the Chicago Section of the American Chemical Society. The purpose of the award is "to publicly recognize eminent chemists who, through years of application and devotion, have brought to the world developments that enable everyone to live more comfortably and to understand the world better." The list of awardees reads like the list of Nobel Prize winners. Paul Flory won this medal in 1973 (Figure 16.5). The first winner was Svante Arrhenius in 1911. Vladimir Ipatieff won in 1940 and Carl Marvel in 1950. Herman F. Mark (1975) and William O. Baker (1978) received this award after Flory.

The Priestley Medal is the highest honor bestowed by the American Chemical Society and is awarded for distinguished service in the field of chemistry (Figure 16.6). It was established in 1922 and Ira Remsen was the first winner in 1923. Paul Flory received this award in 1974. Edward R. Weidlein, the former director of the Mellon Institute was the awardee for 1948. Carl Marvel (1956) and Henry Eyring (1975) are two other friends of Flory that were also so honored.

1974 was a good year for medals. Paul John Flory received the Nobel Prize in chemistry "for his fundamental achievements, both theoretical and experimental, in the physical chemistry of macromolecules" (Figure 16.7).

The final medal for 1974 was the National Medal of Science (Figure 16.8).

In 1976 the Division of Polymer Chemistry of the American Chemical Society established an award for "outstanding research and leadership in

FIGURE 16.6
Joseph Priestley Medal.

polymer science." The first winner was Paul John Flory. He was followed by Carl Marvel and Maurice Huggins. The award was officially named for Herman F. Mark in 1989.

In 1977 Paul Flory received the Perkin Medal of the Society of the Chemical Industry. The award was presented at a gala dinner in New York, at which I was present. Paul Flory lectured the captains of industry about the importance of fundamental research: discoveries are often made adventitiously,

FIGURE 16.7
Nobel Medal.

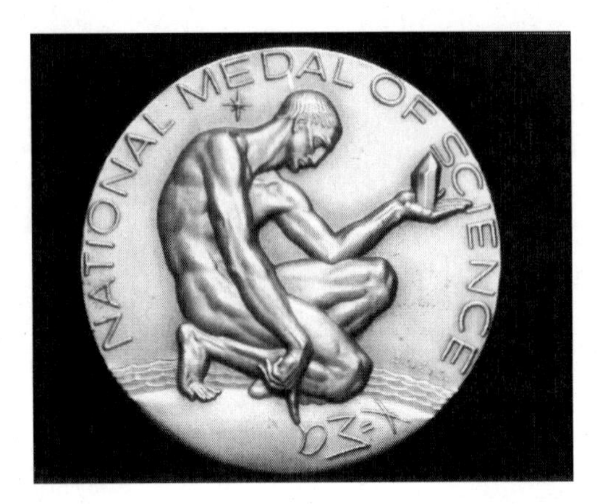

FIGURE 16.8
National Medal of Science.

but they are recognized as discoveries because of their relationship to the known models of pure science.

Paul Flory was appreciated in all countries that manufactured or studied polymers. In 1977 he received the Carl Dietrich Harries Medal of the Deutsche Kautschuk Gesellschaft (German Rubber Society) (Figure 16.9).

After the reception of the Nobel Prize, Paul Flory received many honorary doctorates: University of Akron (1975), Weizmann Institute of Science, Israel (1976) (Figure 16.10), Indiana University (1977), Clarkson College (1978), University of Cincinnati (1982), and University of Massachusetts (1982).

Fittingly, the final major award was given by the Delaware Section of the American Chemical Society in 1985: The Wallace H. Carothers Award.

FIGURE 16.9
Carl Dietrich Harries Medal.

FIGURE 16.10
Doctor of Philosophy Honoris Causa from The Weizmann Institute.

One of the most important aspects of such a collection of awards is that the honoree lends prestige to the awards. Flory was glad to receive such awards, but he was solidly centered on his work and was not influenced personally by the adoration. He joins a select group of highly honored scientists from the twentieth century.

17

Paul John Flory: Humanitarian

Paul Flory lived a life that viewed all scientists as part of the same community. He was a champion for freedom of investigation all over the world. He was willing to use his influence to encourage good scientists wherever they were found. There are four main areas that will be discussed in this chapter: (1) advocacy for specific individuals, (2) group efforts on behalf of human rights, (3) specific opposition to governmental misbehavior, and (4) activities in connection with the National Academy of Sciences and the United States Government.

When Paul Flory met a promising scientist from another country, he was willing to lobby the foreign government to allow them to work with Paul as a postdoctoral fellow. This started well before the fall of the Iron Curtain. Witold Brostow is an example noted above. I remember when Witold arrived from Poland. He was still being "tailed" by a Communist minder. He was careful to defend communism. But, when it was clear that he was truly free, he embraced the United States and freedom. Another scientist mentioned by Burak Erman was Yuri Yarim Agaev from Russia. I remember an occasion at an American Physical Society meeting of the high polymer physics division where every other person at the table for lunch had been assisted by Paul Flory in emigrating to the West. Jan Bares and his wife were present (from Czechoslovakia).

Paul Flory was also active in lobbying governments to allow emigration for humanitarian purposes. One of the best documented cases involves Virgil Percec from Romania (Figure 17.1). He defected from Romania at the International Union of Pure and Applied Chemistry (IUPAC) meeting in Strasbourg, France in 1981. By 1982 he had been appointed as an assistant professor at Case Western Reserve University in Cleveland, but his wife and child were still in Romania. In March he wrote to Paul Flory, at the suggestion of Professor Joseph P. Kennedy, a long-time friend, asking for help in liberating his wife and daughter. Paul Flory was able to assist them in getting the exit visa and they were reunited with Virgil. Percec is now a famous scientist and is the P. Roy Vagelos professor of chemistry at the University of Pennsylvania.

Paul Flory was a member of many human rights organizations, but the one for which he is best-known is "Scientists for Sakharov, Orlov, and Sharansky." The group was founded by Morris Pripstein of Lawrence Berkeley Laboratories (who delivered one of the eulogies at Flory's memorial service). The efforts of the group were largely successful and Pripstein was

FIGURE 17.1
Vergil Percec.

able to greet Andrei Sakharov in 1988 at a meeting of the New York Academy of Sciences (Figure 17.2). Natan Sharansky went on to serve in the Israeli government and Yuri Orlov went to Cornell University. Paul Flory is best known in this context for volunteering to serve as a hostage, so that Yelena Bonner (Sakharov's wife) could come to the West for sorely needed medical care.

Paul Flory was highly opposed to the practice of the Soviet Union of restricting the travel of scientists. When the International Symposium on Macromolecular Chemistry was proposed to meet in Tashkent, USSR, Flory

FIGURE 17.2
Morris Pripstein and Andrei Sakharov at the New York Academy of Sciences in 1988.

FIGURE 17.3
Herbert Morawetz.

joined Herbert Morawetz (Figure 17.3) in organizing a boycott in retaliation. Morawetz and Flory were old friends and had made many contributions to the theory of polymer solutions.

Morawetz escaped the Nazi invasion of Czechoslovakia and emigrated to Canada. He was educated at the Brooklyn Polytechnic Institute and is still a professor of chemistry there. In addition to his many scientific papers, he is the author of the best book on the history of polymer science: *Polymers: The Origin and Growth of a Science* (Wiley, 1985).

As the Tashkent meeting approached, Paul Flory, Nick Tschoegl (Cal Tech), Walter Stockmayer (Dartmouth), and Herman Mark (Brooklyn Poly) sent a letter to Professor V.V. Korshak, chairman of the organizing committee. Among many things, the letter asserted: "Foremost in our minds at this moment is the spectre of the current trials and persecutions of scientists who, according to the evidence available, are guilty of no offense other than expression of their views, and their insistence that the inalienable rights of individuals be respected. The rights in question have been guaranteed by international agreements endorsed by a number of nations, including both the Soviet Union and the United States." This letter led to an international response by leading polymer scientists such as Robert Stepto of the Victoria University of Manchester and Alex Silberberg of the Weizman Institute. While a few Americans did attend the Tashkent Conference, it was Paul Flory and his international community of polymer scientists for human rights that carried the day.

Paul Flory was not bashful when it came to promoting human rights in the National Academy of Sciences. He was selected, along with the President Philip Handler (Figure 17.4), to be a delegate to the "Helsinki Commission" discussions in Hamburg in 1980. As a Nobel Prize winning scientist, he felt a special burden to speak on behalf of all scientists. William Johnson of Stanford put it this way: "Paul was not the sort of person whose ego was

FIGURE 17.4
Paul Flory and Philip Handler.

inflated by receiving the Nobel Prize. Nevertheless, he was very pleased because the prominence and media interest that the Nobel laureate commanded afforded him the opportunity to be much more effective than before in his work on human rights issues." In his obituary for Flory, Henryk Eisenberg, the famous Israeli scientist, noted: "I would like in particular to stress his efforts on behalf of dissident scientists in Eastern Europe, efforts which he pursued with great vigor and devotion. Personalities of his intellectual and moral caliber are indeed few."

18

Paul John Flory: Scientist

Every highly successful scientist has their own approach to the philosophy and practice of science. Flory was no exception. He was convinced that the observable world could be accurately described and cogently explained. Although he is often remembered as a theorist, it would be much better to view him as a complete scientist. He started with a thorough knowledge of the established theories of science. For Paul Flory, the king of theories was thermodynamics. If something violated thermodynamics, it was wrong, not just mistaken! Many of the unfortunates who felt the wrath of Paul Flory violated his first law: the second law of thermodynamics is always right!

Although he rarely made use of quantum mechanics, he believed that the full range of theory should be known and used when appropriate. He was especially known for his insightful use of classical statistical mechanics. I will present several detailed cases involving his work in this area below. He was also highly knowledgeable in the area of chemical kinetics. Henry Eyring (Figure 18.1) was a good friend and Flory was known and admired by this community. When I was a graduate student at Stanford, Paul Flory opened an office in his laboratory for Henry Eyring so that he would be able to work effectively on science while he was being treated for cancer at the Stanford Medical School. Eyring and I spent many hours together and worked on several research projects. Like Paul Flory, he never stopped learning about new things. One of our projects involved mycoplasmas!

Paul Flory also believed that unless the hard work needed to fully describe a physical phenomenon was carried out, there was no point in theorizing. He always insisted that, while there needed to be an overall context for experiments, no detailed theory should be attempted until the experimental situation was settled. Many of the controversies in which he was involved resulted from shoddy experimental work followed by nonsensical theorizing on the part of others. Flory believed that when a physical system was understood, all experiments pointed in the same direction. Isolated artifacts coupled with ad hoc explanations were not real science to Paul Flory. He drilled this concept into me!

While most of his own work was theoretical, he was a great advisor for experimentalists. He suggested experiments that were actually worth doing! There was a sufficient theoretical basis to suggest interesting results. The technology necessary to carry out the experiments was either established or just coming on line. When I joined his group, the visible light laser was just becoming commercially available. This made light scattering a newly

FIGURE 18.1
Henry Eyring (1901–1981). (Adapted from www.patheos.com.)

exciting field. One of my first tasks in the Flory laboratory was to rebuild the SOFICA light scattering photometer that employed a high-pressure mercury arc lamp as the light source. Lasers were very stable with regard to intensity, unlike mercury arcs. While I did restore the classic SOFICA to daily use, I was glad that I did not need to use it myself! Whenever an experimental technique was reliable enough to provide new insight into a problem of interest to Flory, he reached out to a group doing this work. He was a friend of all true experimentalists: make sure your results are so reliable that no one will ever need to correct them until there are better instruments available. He admired Michael Faraday for exactly this reason.

When the experimental situation was as sure as it could be at the time, and the basic theory of the experiment was understood enough to draw reliable conclusions, then Flory was ready to theorize. For example, the distribution of molecular weights for condensation polymers was established as an experimental result. Why was the mathematical most probable distribution an empirically adequate representation of this system? Polymers are formed in a chemical system that starts from monomers and can both add and depolymerize until equilibrium is reached. Both thermodynamics and the statistics of equilibrium are combined to produce the distribution as a function of the extent of reaction. Experiments; theory; insight; success! This paradigm continued to be successful for Flory as he treated different chemical systems and even gels. When all aspects of the program make sense, provisional understanding results. This does not mean that all possible knowledge is obtained, or that the theory is true under all circumstances. It is not necessary to have a theory of everything in order to gain useful understanding. True understanding can be comprehended at many levels of description. A completely correct theory that cannot be applied to any actual system is of no industrial use, or even any use to an experimentalist. Paul Flory understood the value of useful theory!

The theory of liquid mixtures was in its infancy when Flory, and independently Maurice Huggins, began to consider new statistical models of mixing. It was known that two component liquid mixtures could have an excess volume of mixing that was either positive or negative. It was known that the excess heat of mixing could also be of either sign. The notion of an "ideal liquid mixture" was formulated as a starting point. Both Flory and Huggins discovered that the traditional ideal liquid mixture theory was sadly in error. The molar volumes of the two components were usually ignored, and this had a profound effect on the calculated entropy of mixing. For mixtures involving polymers, the effect was quite large. Why is it that the Flory–Huggins ideal mixing is still not the standard result in physical chemistry texts?! Bad traditions die hard.

Individual linear chain molecules swell in solution, and the amount can be either positive or negative. Flory knew the experimental results well. He knew that a chain molecule can be treated as an elastic system, both in extension and in compression. He was able to solve both the problem of gel swelling, and single molecule swelling because he had insight into the structure of the mixture, and into the statistics of the molecules. Peter Debye had an intuition that single molecule swelling would reach an asymptotic value since chain segments far separated on the molecule would cease to be able to interact. Paul Flory believed that the statistical theory of polymer chains was a good description and that allowing the statistical mechanical theory to make the predictions was better than a gut feeling. His theory that the single chain expansion factor continued to increase as the 0.1 power of the chain length indefinitely outraged Debye and he denounced Flory at a seminar at Cornell. In time, Debye came to realize that good theory can be trusted to produce reliable results in the hands of a good theorist. Flory was patient, and in 1953 he was elected to the National Academy of Sciences.

Chemists love to visualize chemical systems, but two dimensions are very different than three dimensions when it comes to condensed phase systems. Paul Flory had a profound geometric sense. He also had the faith to believe that even polymers could achieve the true liquid state. Many controversies involved attempts by outdated workers to impose discredited ideas on the developing world of polymer science. Herman Staudinger was much more influenced by the Kolloid King, Wolfgang Ostwald, than by the emerging field of physical chemistry, the world of Wilhelm Ostwald. Wolfgang hoped to create a new science with new principles of action, new vocabulary, and new heroes. Paul Flory hoped to develop polymer science as an integral part of molecular chemistry. Many European scientists refused to believe that liquids were anything but polycrystalline structures. They were always looking for "order in polymers," even when there was no experimental evidence to support their preordained conclusions.

Paul Flory always started with both the known experimental results and an appropriate model of the physical system. While Staudinger insisted that polymer molecules were "straight," just like they appeared on "paper," Flory believed in a statistical ensemble of chain conformations. Even so, how could such molecules pack into a liquid with no solvent? One of the earliest results of x-ray scattering from natural rubber showed that in the relaxed state, the scattering pattern was consistent with a random liquid. Stretching produced crystallinity and the associated spot pattern. Apparently some polymer molecules were flexible enough to pack into a liquid state. This conclusion was offensive to the highly entrenched forces of European physics. American disciples eagerly sought experimental evidence to refute the notion that there were polymeric liquids. One technique that was brought forward in the 1970s was electron microscopy. Gregory Yeh published micrographs of amorphous solid polymeric films that were not completely uniform in texture. Certainly these "pictures" were proof of "nodules" in the system. Edwin L. Thomas and Donald Uhlmann of MIT (Massachusetts Institute of technology) actually understood electron microscopy and knew that electrons were quantum mechanical objects, just like photons. He showed that a completely random system would produce a "speckle pattern" in the electron microscope, just like a completely amorphous glass produces a speckle pattern in the visible microscope. Paul Flory and I participated in a Symposium on "Order in the Amorphous State" at the Atlantic City ACS Meeting in 1973. By this point the controversy was officially over, and Yeh did not attend. But, Flory knew 40 years before that dynamic amorphous systems can achieve the liquid state if there is enough flexibility.

In any chain molecule, there is order along the chain. In any condensed amorphous system, there is positional order imposed by the packing. Even liquid argon has an interesting scattering function $S(q)$ corresponding to the pair-correlation function for the atoms, $g(r)$. Flory never stopped until he was able to describe the x-ray or neutron scattering from liquid polymers. Do Yoon played an important part in this success.

The success of the Flory–Huggins theory implied that both the single component liquids and the liquid mixtures were random. However, not all liquids and not all mixtures obeyed this result. What additional factor needed to be taken into account? Another measure of the local structure of a liquid is the orientation correlation function $g_2(r)$. For a completely randomly oriented liquid, the correlation is 1. As an isotropic liquid approaches the nematic state, the value of g_2 increases as $(T/(T - T_c)$, where the isotropic–nematic transition temperature for a fully second-order phase transition is denoted T_c. The observed transition is actually first order and occurs slightly above the ideal temperature. *n*-Alkane liquids display this temperature dependence in the pure liquid state. Dilution with an isotropic solvent like carbon tetrachloride can attenuate this effect. Actual examples of polymeric liquids that undergo a true isotropic–nematic phase transition are now common. Flory was already thinking in these terms in 1954 during his visit to the

Victoria University of Manchester. He always tried to include the factors that were necessary in order to comprehend a particular polymeric system.

When some of Paul Flory's most famous theories are expressed in very simple form, such as his expression for the expansion coefficient due to the excluded volume effect in dilute solution, it is important to cite some of his theories that were extremely complicated and were at the research edge in theory, even mathematically. The modern theory of liquids requires an expression for the pair potential of mean force, $u(r)$. Flory and Krigbaum treated a dilute polymer solution as a collection of centers interacting by such a potential. With all the pieces in place, it was possible to write a theory of light scattering from such a solution, and Flory and F. Bueche did just that. The experimental realization of this result needed to wait until all the components could be assembled, but the theory agreed with the scattering function until the point where the chains began to overlap. This required solving integral equations and other highly computationally intensive procedures. Flory was never deterred by the level of mathematics required to make progress, once the experimental situation demanded it.

Another area of mathematical sophistication involved the applications of the rotational isomeric state (RIS) model to the calculation of averaged properties of real chain molecules. This involved matrix methods that taxed the capacities of the computers available in the 1970s. The RIS calculation of the scattering function for depolarized scattering from chain molecules required 147×147 matrices. If the calculations were possible, Paul Flory was eager to make them happen!

While Paul Flory cared about truth in the abstract, he cared even more about achieving useful understanding of physical reality. He was the type of scientist that was willing to do what it takes to arrive at this point. Since he thought deeply, and reached conclusions over a long period of time, he rarely was willing to instantly change his views. But, when new good data became available, and new concepts had been proven, Paul Flory always moved on to the next level of understanding.

19

Paul John Flory: Friend

Throughout this biography many of Paul Flory's friends have appeared. What was it that led them to admire and be drawn to Flory as a friend? His time at Manchester College set the tone. He actively reached out to those with common interests. He was also popular with the general student body, and was elected to positions of responsibility. He made lifelong friendships there and returned many times to Manchester. He more than repaid his debt to the cradle of his scientific career. His friendship with Carl Holl, the chemist and dean, was especially precious to Paul Flory. He admired people who were genuine and energetic.

Since his scientific goal was to gain true understanding, he was quick to reach out to other scientists working in the same area. His career-long friendship with Maurice Huggins is a good example. Walter Stockmayer (Figure 19.1) is another example of someone he met early in his career and with whom he formed a solid relationship. They could count on one another to give their best, both personally and professionally.

While they usually agreed about scientific matters, when there was a disagreement, there was always a firm basis for discussion and a desire to search for better mutual understanding. Stockmayer was known for his gracious manner and friendly spirit, and he was a good influence on Paul Flory.

While Paul Flory was in New Jersey, he was able to interact with people at the Brooklyn Polytechnic Institute, like Herman Mark (1895–1992) (Figure 19.2). In 1985, Mark introduced Flory at his 75th birthday party (Figure 13.6). A long career of profitable scientific interactions characterized their friendship. Flory was admired by the early leaders in polymer science because he brought both rigor and energy to the field.

Paul Flory also became friends with Geoffrey Gee in this era. They were intellectual equals and encouraged one another. They shared a passionate devotion to thermodynamics. Flory was indeed passionate, and he responded warmly to those who shared his passion for science.

The Goodyear period allowed Paul Flory to make a large number of great friends. Carl Marvel (1894–1988) and Fred Wall (1912–2010) from Illinois were regular colleagues at the Rubber Round Table. Since Wall was one of the leaders in thermodynamics and statistical mechanics, they published a landmark paper on the theory of rubber elasticity together. Other close friends from that group included Peter Debye from Cornell and Bill Baker from AT&T Bell Laboratories. Baker consulted with Flory about many aspects of polymer science from the 1940s to his death. He even asked if Paul had any

FIGURE 19.1
Walter Stockmayer (1914–2004). (From Dartmouth College. With permission.)

good students that might like to work there! Debye also remembered Paul Flory when the war had ended and invited him to be the Baker Lecturer. Even in the midst of serious disagreements, Paul Flory retained a warm personal relationship with Debye. He was more committed to a joint search for truth than a personal victory. As a result, Paul Flory was usually on the winning side in most disputes.

FIGURE 19.2
Herman Mark. (From Interscience. With permission.)

The Cornell University Chemistry Department was filled with strong personalities. Paul Flory viewed this as a great virtue. He collaborated with the other physical chemists and great science resulted. They shared students, postdocs, and ideas. Physical chemistry has been the beneficiary. Harold Scheraga is still hard at it! Flory also developed strong bonds with his students and postdoctoral fellows. Leo Mandelkern was perhaps the one alongside of whom Paul Flory fought the most battles. Paul Flory engaged his junior colleagues as scientists. If they were willing to pursue the goal with rigor and energy, he was a life-long champion for them.

The move to the Mellon Institute required a strong administrator to mind the quotidian aspects of the institution. Tom Fox was already established at Rohm and Haas, but gladly joined Paul Flory as assistant director (Figure 19.3). When Paul Flory left the Mellon Institute, Fox stayed on and helped with the transition to Carnegie Mellon University. Fox was the perfect friend. He brought all he had to the task, and worked seamlessly with others. He was Flory's greatest gift to the Mellon Institute. The large number of joint papers reflected their complete confidence in one another. Paul Flory

MELLON INSTITUTE NEWS

Vol. XX Thursday, April 11, 1957 No. 28

DR. THOMAS G. FOX

APPOINTED ASSISTANT DIRECTOR OF RESEARCH

FIGURE 19.3
Thomas G Fox (G was his middle name). (From Carnegie Mellon University Archives. With permission.)

inspired that kind of confidence in others and they produced their best when it was a joint project.

Stanford University proved to be a good place to pursue science and to attract good students and postdoctoral fellows. Bill Johnson and Henry Taube were staunch friends throughout and wrote his eulogy for the National Academy of Sciences (along with Walter Stockmayer). They declared "Paul Flory was a warm and loyal friend to those people who, like he, had high standards of integrity and were honestly modest about their own accomplishments and potential. These friends in turn greatly admired Paul."

Jim Mark was an example of one of Paul Flory's true friends. He was attracted to Paul before he went to Stanford, both through his work at Rohm and Haas and through the University of Pennsylvania. He chose to work within the overall paradigm defined by Flory throughout his career. He organized events at scientific meetings, edited Paul Flory's Collected Papers, and generally served as a goodwill ambassador for Paul Flory and for his science. Flory appreciated this level of support and was a constant source of encouragement to Jim.

Paul Flory was revered in Japan, and he had several outstanding postdoctoral fellows from that country. His closest friend was Akihiro Abe. Aki was proud to be a "Flory-watcher." He never failed to learn something new in his presence. Paul Flory sought to share his knowledge and insights with his friends. Those who chose to listen were truly blessed. Abe was able to experience Paul Flory's true friendship on many occasions and remains a force in polymer science in Japan. For Paul Flory, science was truly a universal activity with no "home country." As a result he was welcomed by scientists throughout the world.

Robert Orwoll was attracted to Paul Flory by his friendliness and modesty. When students responded to this encouragement, a profitable relationship followed. I benefited from the work of Orwoll throughout my own career, since his careful measurements on *n*-alkanes are still the best! Although many of the narratives mentioned "tea-time," there is a reason why it remained such a positive memory: Paul Flory truly cared about his students and sought to provide a relaxed and highly interactive time each day. The only person in the history of the Flory laboratory to reject this daily custom was Henry Eyring. He referred to tea as "footwash" and adhered to his Mormon convictions!

Real friendship is intense. None of Paul Flory's friends were afraid of him, and were spurred on to greater achievement by their strong interactions. Paul Flory was on a mission to achieve coherent understanding. Sloppy thinking or shoddy calculation was not the way forward. Bob Jernigan cites this passion for perfection as one of his most important lessons. Real scientists are never satisfied with half-hearted attempts to understand.

Paul Flory was linked as part of many scientific circles. Leonard M. Peller (1928–2000) is mentioned in Dave Brant's essay, and I remember him well during my time at Stanford. He was an expert in biochemically important phase transitions. Paul Flory was interested in all macromolecules (and all molecules in general). He reached out to many people in the biophysical chemistry community and tried to make sure all his work was consistent with larger communities of interest. Dave Brant learned this lesson well and went on to a fine career in the study of polysaccharides.

One of my colleagues at AT&T Bell Laboratories was Alan Tonelli. His most vivid memory of Paul Flory was his kindness and his flexibility. Flory was very rigorous when it came to the laws of nature, but he was flexible toward the laws of men. I experienced this side of Paul Flory when I faced the Draft Board. All conscientious objectors were asked to perform alternative service. Flory arranged for me to satisfy this requirement at the Stanford Medical Center! While Alan's early career was characterized by uncertainty about the future, Paul Flory ignited a flame that has never gone out. This ability to inspire graduate students to reach their full potential is rare, but fully realized with Flory.

Another student who achieved her full potential is Wilma King Olson. Paul Flory actively promoted her career. He sent her to meetings and arranged for her to meet the important people in her field. He stimulated a postdoctoral fellowship near her husband's new job. And he supported her appointment at Rutgers. She has more than rewarded him by excelling in the field of biopolymers.

Paul Flory had a special fondness for international students. Rather than meet them only in his office or the laboratory, Flory invited them into his home and his heart. All of the narratives dwell on Flory's sense of genuine hospitality. Paul Flory also knew how to receive hospitality. He was at home anywhere in the world (except perhaps in the Kremlin). Burak Erman, like Akihiro Abe, enjoyed entertaining Paul and Emily in their home countries. In Flory's "Republic of Polymer Science," scientists of all countries could pursue a Socratic quest for physical understanding.

Paul John Flory was a true friend to those who joined him in his passionate search for scientific understanding and human justice. The sheer number of his close friends was staggering, especially for someone who loved to "be alone with his thoughts." Because he had such a strong "center," he could reach out with confidence to others. I am truly thankful for his friendship and have tried to bring that same spirit to the world of chemical history. Flory will be remembered in the history of chemistry, but not just as a Nobel Prize winner in chemistry. Had he lived longer, he might have joined Linus Pauling in winning the Nobel Peace Prize as well. May he "rest in peace."

References

1. James, L.K. (Ed.), *Nobel Laureates in Chemistry* 1901–1992, American Chemical Society and the Chemical Heritage Foundation, Philadelphia, 1993.
2. Flory, P.J. *Principles of Polymer Chemistry*, Cornell University Press, Ithaca, NY, 1953.
3. Patterson, G.D., Paul John Flory: Physical chemist and humanitarian. In: Patterson, G.D., and Rasmussen, S.C. (Eds.), *Characters in Chemistry, ACS Symposium Series* 1136, American Chemical Society, Washington, DC, 2013.
4. Patterson, G.D., *Polymer Science* 1935–1953. *Consolidating the Paradigm*, Springer, New York, 2014.
5. Johnson, W.S., Stockmayer, W.H., Taube, H., *Paul John Flory* 1910–1985, *National Academy of Sciences Biographical Memoirs*, Vol. 82, National Academy Press, Washington, DC, 2002.
6. *Complete Dictionary of Scientific Biography*, Vol. 21, 37–41, Charles Scribner's Sons, Detroit, MI, 2008.
7. Brumbaugh, D.F. (Ed.), *The Brethren Encyclopedia*, Brethren Encyclopedia, Inc., Philadelphia, 1983–2005.
8. Flora, J.C., *A Genealogy and History of Descendants of Jacob Flora Senior of Franklin County, Virginia*, Church Center Press, Myerstown, PA, 1951.
9. Brumbaugh, M.G., *A History of the German Baptist Brethren in Europe and America*, Brethren Publishing House, Elgin, IL, 1899.
10. www.ancestry.com "Flory Family Tree", 2014.
11. *Manchester College Bulletin*, Manchester, IN, January 1975.
12. Eberly, W.R., *The Story of the Natural Sciences at Manchester College*, Manchester College, North Manchester, IN, 2005.
13. Flory, P.J., The values of science, Manchester University Archives and Brethren Historical Collection, 1960.
14. Lewis, G.N., Randall, M., *Thermodynamics and the Free Energy of Chemical Substances*, McGraw-Hill, New York, 1923.
15. Adam, N.K., *Physical Chemistry*, Oxford University Press, Oxford, 1930.
16. Hinshelwood, C.N., *The Kinetics of Chemical Change in Gaseous Systems*, Oxford University Press, Oxford, 1926.
17. Hounshell, D.A., Smith, J.K., *Science and Corporate Strategy: DuPont R&D* 1902–1980, Cambridge University Press, Cambridge, 1988.
18. Patterson, G.D., *A Prehistory of Polymer Science*, Springer, New York, 2012.
19. Flory, P.J., The mechanism of vinyl polymerization, *J. Am. Chem. Soc.*, 59, 241, 1937.
20. Flory, P.J., Overberger, C.G., Reflections by an eminent chemist: Dr. Paul J. Flory. In: *Eminent Chemist Video Tape Series*, American Chemical Society, Washington, DC, 1982.
21. Flory, P.J., *Reflections by Two Eminent Chemists, Drs. H. Mark and P. J. Flory*, American Chemical Society, Washington, DC, 1982.
22. Whissel, P., *The First Cooperative College: A History of the College of Engineering at the University of Cincinnati*, University of Cincinnati, Cincinnati, OH, 1993.

23. Flory, P.J., Intramolecular reaction between neighboring substituents of vinyl polymers, *J. Am. Chem. Soc.*, 61(6), 1518–1521, 1939.
24. Flory, P.J., Kinetics of polyesterification: A study of the effects of molecular weight and viscosity on reaction rate. *J. Am. Chem. Soc.*, 61(12), 3334–3340, 1939.
25. Flory, P.J., Viscosities of linear polyesters: An exact relationship between viscosity and chain length, *J. Am. Chem. Soc.*, 62(5), 1057–1070, 1940.
26. Flory, P.J., Molecular size distribution in ethylene oxide polymers, *J. Am. Chem. Soc.*, 62(6), 1561–1565, 1940.
27. Flory, P.J., Kinetics of the degradation of polyesters by alcohols, *J. Am. Chem. Soc.*, 62(9), 2255–2261, 1940.
28. Flory, P.J., A comparison of esterification and ester interchange kinetics, *J. Am. Chem. Soc.*, 62(11), 2261–2264, 1940.
29. Flory, P.J., Stickney, P.B., Viscosites of polyester solutions and the Staudinger equation, *J. Am. Chem. Soc.*, 62(11), 3032–3038, 1940.
30. Flory, P.J., Molecular size distribution in three dimensional polymers: I, Gelation, *J. Am. Chem. Soc.*, 63, 3083–3090, 1941.
31. Flory, P.J., Molecular size distribution in linear condensation polymers, *J. Am. Chem. Soc.*, 58(10), 1877–1885, 1936.
32. Flory, P.J., Constitution of three-dimensional polymers and the theory of gelation, *J. Phys. Chem.*, 46, 132–140, 1942.
33. Stockmayer, W.H., Theory of molecular size distribution and gel formation in branched polymers, *J. Chem. Phys.*, 12, 125, 1944.
34. Carothers, W.H., Polymers and polyfunctionality, *Faraday Soc. Trans.*, 32, 39–53, 1935.
35. (a) Flory, P.J., Molecular size distribution in three dimensional polymers: II, Trifunctional branching units, *J. Am. Chem. Soc.*, 63(11), 3091–3096, 1941; (b) Flory, P.J., Molecular size distribution in three dimensional polymers: III, Tetrafunctional branching units, *J. Am. Chem. Soc.*, 63(11), 3096–3100, 1941.
36. Flory, P.J., Fundamental principles of condensation polymerization, *Chem. Rev.*, 39(1), 137–197, 1946.
37. Flory, P.J., Thermodynamics of high polymer solutions, *J. Chem. Phys.*, 9, 660, 1941.
38. Flory, P.J., Thermodynamics of high polymer solutions, *J. Chem. Phys.*, 10, 51, 1942.
39. Flory, P.J., Rehner, J. Jr., Statistical mechanics of cross-linked polymer networks: I, rubberlike elasticity, *J. Chem. Phys.*, 11, 512, 1943.
40. Flory, P.J., Rehner, J. Jr., Statistical mechanics of cross-linked polymer networks: II, swelling, *J. Chem. Phys.*, 11, 521, 1943.
41. Patterson, G., *A Prehistory of Polymer Science*, Springer, New York, 2012.
42. Morris, P.J.T., *The American Synthetic Rubber Research Program*, University of Pennsylvania Press, Philadelphia, 1989.
43. Flory, P.J., Network structure and the elastic properties of vulcanized rubber, *Chem. Rev.*, 35, 51, 1944.
44. Flory, P.J., Fundamental principles of condensation polymerization, *Chem. Rev.*, 39, 137, 1946.
45. Flory, P.J., Thermodynamics of dilute solutions of high polymers, *J. Chem. Phys.*, 13, 453, 1945.
46. Flory, P.J., Thermodynamics of crystallization in high polymers: I, crystallization induced by stretching, *J. Chem. Phys.*, 15, 397, 1947.

47. Flory, P.J. Internal Goodyear Memorandum to Dr. L.B. Sebrell, Hydrocarbon polymers for non-rubber uses, Goodyear Archives at the University of Akron, April 3, 1944.
48. Schaefgen, J.R., *Oral History*, Chemical Heritage Foundation, Philadelphia, 1986.
49. McMillan, F.M., *The Chain Straighteners*, The Macmillan Press, London, 1979.
50. Bishop, M., *A History of Cornell*, Cornell University Press, Ithaca, NY, 1962.
51. Shultz, A.R., Flory, P.J., Phase equilibria in polymer-solvent systems, *J. Am. Chem. Soc.*, 74, 4760, 1952.
52. Fox, T.G. Jr., Flory, P.J., Intrinsic-viscosity-molecular weight relationships for polyisobutylene, *J. Phys. Colloid Chem.*, 53, 197, 1949.
53. Fox, T.G. Jr., Flory, P.J., Further studies on the melt viscosity of polyisobutylene, *J. Phys. Colloid Chem.*, 55, 221, 1951.
54. Fox, T.G. Jr., Flory, P.J., Second-order transition temperatures and related properties of polystyrene: I, influence of molecular weight, *J. Appl. Phys.*, 21, 581, 1950.
55. Flory, P.J., Fox, T.G Jr., Treatment of intrinsic viscosities, *J. Am. Chem. Soc.*, 73, 1904, 1951.
56. Fox, T.G. Jr., Flory, P.J., Bueche, A.M., Treatment of osmotic and light scattering data for dilute solutions, *J. Am. Chem. Soc.*, 73, 285, 1951.
57. Mandelkern, L., Flory, P.J., The dependence of the diffusion coefficient on concentration in dilute solution, *J. Chem. Phys.*, 19, 984, 1951.
58. Flory, P.J., *Principles of Polymer Chemistry*, Cornell University Press, Ithaca, NY, 1953.
59. Allen, G., Geoffrey Gee, C.B.E. 6 June 1910–13 December 1996, *Biogr Mem Fellows R Soc*, 4, 184, 1999.
60. University of Manchester Council Reports, University of Manchester Archives.
61. Flory, P.J., Statistical thermodynamics of semi-flexible chain molecules, *Proc. R. Soc. London*, A234, 60, 1956.
62. Flory, P.J., Phase equilibria in solutions of rod-like particles, *Proc. R. Soc. London*, A234, 73, 1956.
63. Mellon Institute: 50 Years of Progress, Pittsburgh, PA, 1963.
64. Hutchinson, E., The department of chemistry Stanford University 1891–1976: A brief account of the first eighty-five years, 1977.
65. Flory, P.J., On the morphology of the crystalline state in polymers, *J. Am. Chem. Soc.*, 84, 2857, 1962.
66. Mark, J.E., Flory, P.J., Stress-temperature coefficients for isotactic and atactic poly-(butane-1), *J. Phys. Chem.*, 67, 1396, 1963.
67. Mark, J.E., Flory, P.J., Configuration of the poly-(dimethylsiloxane) chain: I, the temperature coefficient of the unperturbed dimensions, *J. Am. Chem. Soc.*, 86, 138, 1964.
68. Mark, J.E., Flory, P.J., The configuration of the polyoxyethylene chain, *J. Am. Chem. Soc.*, 87, 1415, 1965.
69. Flory, P.J., Mark J.E., The configuration of the polyoxymethylene chain, *Makromol. Chem.*, 75, 11, 1964.
70. Flory, P.J., Abe, A., Thermodynamic properties of nonpolar mixtures of small molecules, *J. Am. Chem. Soc.*, 86, 3563, 1964.
71. Flory, P.J., Orwoll, R.A., Vrij, A., Statistical thermodynamics of chain molecule liquids, I and II, *J. Am. Chem. Soc.*, 86, 3507, 3515, 1964.
72. Flory, P.J., Jernigan, R.L., Second and fourth moments of chain molecules, *J. Chem. Phys.*, 42, 3509, 1965.

73. Flory, P.J., *Statistical Mechanics of Chain Molecules*, Interscience, New York, 1969.
74. Flory, P.J., Yoon, D.Y., Moments and distribution functions for polymer chains of finite length, *J. Chem. Phys.*, 61, 5358, 5366, 1974.
75. Yoon, D.Y., Flory, P.J., Small-angle neutron and x-ray scattering by poly(methylmethacrylate) chains, *Polymer*, 16, 645, 1975.
76. Yoon, D.Y., Flory, P.J., Small-angle x-ray and neutron scattering by polymethylene, polyoxyethylene and polystyrene chains, *Macromolecules*, 9, 294, 1976.
77. Suter, U.W., Flory, P.J., Conformational energy and configurational statistics of polypropylene, *Macromolecules*, 8, 765, 1975.
78. Flory, P.J., Suter, U.W., Mutter, M., Macrocyclization equilibria, *J. Am. Chem. Soc.*, 98, 5733, 5740, 5745, 1976.
79. Erman, B., Flory, P.J., Theory of elasticity of polymer networks: II, The effect of geometric constraints on junctions, *J. Chem. Phys.*, 68, 5363, 1978.
80. Flory, P.J., Ronca, G., Theory of systems of rodlike particles, *Mol. Cryst. Liq. Cryst.*, 54, 289, 311, 1979.
81. Dill, K.A., Flory, P.J., Molecular organization in micelles and vesicles, *Proc. Natl. Acad. Sci. USA*, 78, 676, 1981.
82. Morton, M., Mark, J.E., Paul John Flory (1910–1985), *Rubber Chem. Technol.*, 60, G47, 1987.
83. Rutgers, A.J., *Physical Chemistry*, Interscience Publishers, New York, 1954.
84. Verwey, E.J.W, Overbeek, J.Th.G., *Theory of the Stability of Lyophobic Colloids*, Elsevier, NY, 1948, Dover Publications, Mineola, New York, 1999.
85. Vrij, A., Light scattering from charged colloidal particles in salt solutions, PhD thesis, Utrecht, 1959. (Lichtverstrooiing door Geladen Colloidale Deeltjes in Zoutoplossingen).
86. Hesselink, F.Th, On the theory of the stabilization of colloidal dispersions by adsorbed macromolecules, PhD thesis, Utrecht, 1971.
87. Eisenberg, H., Felsenfeld, G., Studies of the temperature-dependent conformation and phase separation of polyriboadenylic acid solutions at neutral pH, *J. Mol. Biol.*, 30(1), 17–37, 1967.
88. Flory, P.J., Configurational statistics of polypeptide chains. In: Ramachandran, G.N. (Ed.), *Conformation of Biopolymers I*, Academic Press, New York, pp. 339–363, 1967.
89. Sasisekharan, V., Lakshminarayanan, A.V., Ramachandran, G.N. Stèreochemistry of nucleic acids and polynucleotides. I. Theoretical determination of the allowed conformations of the monomer unit. In: Ramachandran, G.N. (Ed.), *Conformation of Biopolymers II*, Academic Press, New York, pp. 641–654, 1967.
90. McGlashan, M.L., *J. Chem. Educ.*, 43, 226, 1966.
91. Orwoll, R.A., Flory, P.J., *J. Am. Chem. Soc.*, 89, 5814, 1967.
92. Höcker, H., Blake, G.J., Flory, P.J., *Trans. Faraday Soc.*, 67, 2251, 1971.
93. Brostow, W., *Macromolecules*, 4, 742, 1971.
94. Brostow, W., Sochanski, J.S., *J. Mater. Sci.*, 10, 2134, 1975.
95. Flory, P.J., *J. Am. Chem. Soc.*, 84, 2857, 1962.
96. Flory, P.J., Yoon, D.Y., *Nature*, 272, 226, 1978.
97. Flory, P.J., Selected papers of Paul John Flory. In: Mandelkern, L., Mark, J.E., Suter, U.W., Yoon, D.Y. (Eds.), *Selected Works*, Vols. I–III, Stanford University Press, Stanford, 1985.
98. Brostow, W., Paul J. Flory: 19 June 1910–8 September 1985, *Mater. Chem. Phys.*, 14, 305, 1986.

Index of Names

Subject Index

T

U

V

W